Selected Titles in This Series

Attendees of the 1st Annual
DNA Based Computer Workshop

DIMACS

Series in Discrete Mathematics and Theoretical Computer Science

Volume 27

DNA Based Computers

Proceedings of a DIMACS Workshop
April 4, 1995
Princeton University

Richard J. Lipton
Eric B. Baum
Editors

NSF Science and Technology Center
in Discrete Mathematics and Theoretical Computer Science
A consortium of Rutgers University, Princeton University,
AT&T Bell Labs, Bellcore

American Mathematical Society

This DIMACS volume on DNA based computers contains papers selected from a workshop held April 4, 1995, at Princeton University as part of the DIMACS Special Year on Mathematical Support for Molecular Biology.

1991 *Mathematics Subject Classification.* Primary 68Q05, 68Q10, 68Q15, 68Q22, 68Q42, 68Q80.

Library of Congress Cataloging-in-Publication Data
DNA based computers: DIMACS workshop, April 4, 1995 / Richard J. Lipton, Eric B. Baum, editors.
p. cm. — (DIMACS series in discrete mathematics and theoretical computer science; v. 27)
"NSF Science and Technology Center in Discrete Mathematics and Theoretical Computer Science, a consortium of Rutgers University, Princeton University, AT&T Bell Labs, Bellcore."
Includes bibliographical references.
ISBN 0-8218-0518-5
1. Molecular computers—Congresses.   I. Lipton, Richard J.   II. Baum, Eric B., 1957– .
III. NSF Science and Technology Center in Discrete Mathematics and Theoretical Computer Science.   IV. DIMACS (Group)   V. Series.
QA76.887.D53   1996
511.3—dc20
96-8191
CIP

Contents

Foreword

This DIMACS volume on DNA based computing contains refereed papers selected from a workshop held at Princeton University as part of the DIMACS Special Year on Mathematical Support for Molecular Biology. We would particularly like to thank the organizers of the program, Richard Lipton and Eric Baum, for their energy and vision to bring together a group of researchers to explore the potential of DNA based computing at such a timely moment.

The workshop and special year programs were made possible by funding from the National Science Foundation, the New Jersey Commission on Science and Technology, AT&T Bell Laboratories, and Bellcore. We gratefully acknowledge their support.

Fred S. Roberts
Director, DIMACS

Andrew Yao,
CoDirector, DIMACS

Stephen R. Mahaney
Associate Director, DIMACS

Introduction

This volume constitutes the proceedings of the conference held April 4, 1995, at Princeton University. The subject of the meeting was the new area of DNA based computing.

The conference was sponsored by DIMACS and NSF. We thank them for their generous support.

The area of DNA based computing is the study of using DNA strands as individual computers. It was started by Len Adleman's initial paper in *Science* in November 1994.

Eric Baum, NEC

Richard J. Lipton, Princeton University

Professor Leonard M. Adleman

DIMACS Series in Discrete Mathematics
and Theoretical Computer Science
Volume **27**, 1996

On Constructing A Molecular Computer [1]

Leonard M. Adleman[2]
Department of Computer Science
University of Southern California
Los Angeles, California 90089, USA
`adleman@cs.usc.edu`

Abstract

It has recently been suggested that under some circumstances computers based on molecular interactions may be a viable alternative to computers based on electronics. Here, some practical aspects of constructing a molecular computer are considered.

1 Introduction

In [1] a small instance of the so called 'Hamiltonian path problem' was encoded in molecules of DNA and solved in a test tube using standard methods of molecular biology. It was asserted that for certain problems, molecular computers might compete with electronic computers. At the time that [1] appeared, there seemed to be formidable obstructions to creating a practical molecular computer. Roughly, these obstructions were of two types:

- Physical obstructions arising primarily from difficulties in dealing with large scale systems and in coping with errors.

- Logical obstructions concerning the versatility of molecular computers and their capacity to efficiently accommodate a wide variety of computational problems.

---

[1]Research in molecular computation is embryonic. This is an informal, speculative 'work in progress' which refelects the author's impressions as of January 1995.

[2]Research supported by the National Science Foundation under grant CCR-9214671.

Happily, there has been rapid and significant progress toward overcoming these obstructions. Lipton [5] has directly addressed the question of versatility by demonstrating a large class of computational problems which are apparently amenable to molecular computation. Further, Lipton's work has simplified the underlying theory to such an extent that it may now be possible to give a clearer picture of how to handle the physical obstructions.

This note gives one (incomplete) view of how a molecular computer might be constructed. The goal is to take a small step in the direction of practicality in the hopes that other researchers will go far beyond.

2   Abstract models

In this section abstract models of molecular computers are described. In the sections which follow physical incarnations of some of these models are given. The reader most interested in practical issues may wish to go immediately to *An Example Application:* in section 3.

We begin with the model described in [5].

The Unrestricted (DNA) Model:

A (test) tube is a set of molecules of DNA (i.e. a multi-set of finite strings over the alphabet $\{A, C, G, T\}$). Given a tube, one can perform the following operations:

1. *Separate.*[3]   Given a tube T and a string of symbols S over $\{A, C, G, T\}$, produce two tubes $+(T, S)$ and $-(T, S)$, where $+(T, S)$ is all of the molecules of DNA in T which contain the consecutive subsequence S and $-(T, S)$ is all of the molecules of DNA in T which do not contain the consecutive subsequence S.

2. *Merge.* Given tubes T_1, T_2, produce $\cup(T_1, T_2)$ where:

$$\cup(T_1, T_2) = T_1 \cup T_2$$

   Throughout this paper $\cup$ denotes the multiset union.

3. *Detect.* Given a tube T, say 'yes' if T contains at least one DNA molecule, and say 'no' if it contains none.

---

[3]Lipton uses the term Extract

4. *Amplify.* Given a tube T produce two tubes $T'(T)$ and $T''(T)$ such that

$$T = T'(T) = T''(T).$$

These operations are then used to write 'programs' which receives a tube as input and return as output either 'yes' , 'no' or a set of tubes. For example, consider the following program:

(1) Input(T)

(2) $T_1 = -(T, C)$

(3) $T_2 = -(T_1, G)$

(4) $T_3 = -(T_2, T)$

(5) Output(Detect(T_3))

On input a tube, it returns 'yes' if the input tube contains a DNA molecule which is composed entirely of A and otherwise returns 'no'. If step (5) is changed to OUTPUT(T_3) then it returns the tube containing exactly those DNA molecules from the input tube which are composed entirely of A.

We next describe a restricted and modified version of the unrestricted model. It is likely that some computational problems which can be efficiently handled using the unrestricted model cannot be efficiently handled using the restricted model. However, it is hoped that the restricted model will be more easily implemented.

First, despite its use in [1], the *Amplify* operation is a concern. Amplification is a complex and rare process. Currently, it can be only be applied to certain special biological molecules (e.g. DNA and RNA) and some living things (e.g. *E. coli.*, phages). It involves the construction of covalent bonds, the consumption of materials and may be prone to error. For the purposes of a 'practical' molecular computer, it may be preferable to avoid it (or restrict its use). Consequently, in what follows *Amplify* will not be used.

Second, it may be preferable to use molecules other than DNA, so we will start with an alphabet Σ which is not necessarily $\{A, C, G, T\}$. Further, though DNA has a natural structure which allows one to order the occurrence of symbols (5' to 3') and hence speak of 'sequences', this may not be true for other types of molecules where the groups encoding the symbols may simply be placed anywhere on a standard (potentially non linear) molecular backbone. Hence elements of a tube will no longer be sequences of symbols from Σ, but rather subsets of symbols from Σ. Such a subset will be called an 'aggregate' . Finally, separation will be allowed with respect to symbols only.

The Restricted Model:

A tube is a multi-set of aggregates over an alphabet Σ. Given a tube, one can perform the following operations:

1. *Separate.* Given a tube T and a symbol $s \in \Sigma$, produce two tubes $+(T, s)$ and $-(T, s)$ where $+(T, s)$ is all of the aggregates of T which contain the symbol s and $-(T, s)$ is all of the aggregates of T which do not contain the symbol s.

2. *Merge.* Given tubes T_1, T_2, produce $\cup(T_1, T_2)$ where:

$$\cup(T_1, T_2) = T_1 \cup T_2$$

3. *Detect.* Given a tube T, say 'yes' if T contains at least one aggregate and say 'no' if it contains none.

Despite the restrictions, there are still problems of apparent interest which might be efficiently solved with the restricted model. For example, consider the well know '3- colorability problem'. The '3-colorability problem' is to decide of an undirected (i.e. with 'two-way' edges) graph $G =< V, E >$ whether each vertex in the graph can be colored either red, blue, or green in such a way that after coloring, no two vertices which are connected by an edge have the same color. The problem is NP-complete.

Given an n vertex graph G with edges $e_1, e_2, ..., e_z$, let
$\Sigma = \{r_1, b_1, g_1, r_2, b_2, g_2, ..., r_n, b_n, g_n\}$.
and consider the following restricted program on input:

$$T = \{\alpha | \alpha \subseteq \Sigma \ \&$$
$$\alpha = \{c_1 c_2, ..., c_n\} \ \& \ [c_i = r_i \text{ or } c_i = b_i \text{ or } c_i = g_i], i = 1, 2, ..., n\}$$

(1) Input(T).
(2) for $k = 1$ to z. Let $e_k =< i, j >$:
      (a) $T_{red} = +(T, r_i)$ and $T_{blue \ or \ green} = -(T, r_i)$.
      (b) $T_{blue} = +(T_{blue \ or \ green}, b_i)$ and $T_{green} = -(T_{blue \ or \ green}, b_i)$.

(c) $T_{red}^{good} = -(T_{red}, r_j)$.
(d) $T_{blue}^{good} = -(T_{blue}, b_j)$.
(e) $T_{green}^{good} = -(T_{green}, g_j)$.
(f) $T' = \cup(T_{red}^{good}, T_{blue}^{good})$.
(g) $T = \cup(T_{green}^{good}, T')$.

(3) Output(Detect(T)).

Clearly the elements of the input tube T are in one to one correspondence with the possible ways to assign colors to the vertices of G. Step (2) of the program creates from the current tube a new tube from which all aggregates α with $c_i = c_j$ (i.e. which assign the same color to vertex i and j) have been removed. After $5k$ separations, $2k$ merges and 1 detect, the program will output 'yes' if G is 3-colorable and 'no' otherwise. Hence the restricted program answers the 3-colorability problem for the graph G in 'linear' time (in the number of edges) while an electronic compute might require exponential time. This is the value of parallelism, each steps act on as many as 3^n inputs simultaneously.

Lipton [5] gives a method for efficiently solving the well known (NP-complete) 'satisfiability problem': given a propositional formula (not necessarily in conjunctive normal form) decide if it is satisfiable. In fact, given a propositional formula ϕ in n variables and m binary connectives (and's or or's) and any number of not's, Lipton's method will after $m+1$ separations and m merges produce two tubes, the first contains molecules encoding satisfying truth assignments, the second contains molecules encoding unsatisfying truth assignments. A single final detect on the first tube determines whether ϕ is satisfiable or not.

For most of the incarnations of restricted computation which are considered here, merging is done by pouring the contents of the input tubes into a single tube. This is presumably much faster and less error prone than separating. For many of the problems attacked by restricted computers, in appears that detect is done rarely. Accordingly, when considering a proposed incarnation, the following parameters will be considered:

Time: The time τ required for a separation.

Accuracy: The error rate of a separation. In a real system separation will not be perfect, there will be errors of inclusion and exclusion. Let ϵ_+ denote the probability that after a separation, an aggregate which should end up in $+(T, s)$ does not. Let ϵ_- denote the probability that after a separation, an aggregate which should end up in $-(T, s)$ does not.

It is of course possible that in real systems the time and accuracy of a separation may depending on the tube and symbol being considered. However, for convenience it will be assumed that a uniform time and accuracy can be given.

The restricted model of molecular computation is memory-less in the sense that the molecules themselves do not change in the course of a computation. The state of the computation is represented primarily by the distribution of the molecules in the various tubes. Next an abstract model of a molecular computer with memory is presented. This model has the same power as the unrestricted model. It is presented for its potential to be realized in a practical way; however, it will be given only passing attention in what follows and can be safely omitted by the reader more interested in practical matters.

The Memory Model:

A tube is a multi-set of aggregates over an alphabet $\Sigma = \Sigma_a \cup \Sigma_b$ where $\Sigma_a = \{a_1, a_2, ..., a_n\}$ and $\Sigma_b = \{b_1, b_2, ..., b_n\}$. Given a tube, one can perform the following operations:

1. *Separate* Given a tube T and a symbol $s \in \Sigma$, produce two tubes $+(T, s)$ and $-(T, s)$ where $+(T, s)$ is all of aggregates of T which contain the symbol s and $-(T, s)$ is all of the aggregates of T which do not contain the symbol s.

2. *Merge.* Given tubes T_1, T_2, produce $\cup(T_1, T_2)$ where:

$$\cup(T_1, T_2) = T_1 \cup T_2$$

3. *Detect.* Given a tube T, say 'yes' if T contains at least one aggregate and say 'no' if it contains none.

4. *Flip.*

   (a) Given a tube T and a symbol $a_i \in \Sigma$, produce a new tube $T^{a_i} = \{\alpha^{a_i} | \alpha \in T\}$ where for all $\alpha \subset \Sigma$, if $a_i \notin \alpha$ then $\alpha^{a_i} = \alpha$ and if $a_i \in \alpha$ then $\alpha^{a_i} = (\alpha - \{a_i\}) \cup \{b_i\}$.

   (b) Given a tube T and a symbol $b_i \in \Sigma$, produce a new tube $T^{b_i} = \{\alpha^{b_i} | \alpha \in T\}$ where for all $\alpha \subset \Sigma$, if $b_i \notin \alpha$ then $\alpha^{b_i} = \alpha$ and if $b_i \in \alpha$ then $\alpha^{b_i} = (\alpha - \{b_i\}) \cup \{a_i\}$.

Hence the flip operations switches a_i to b_i (b_i to a_i) in all aggregates that contain an a_i (b_i).

Let

$$T = \{\alpha | \alpha \subseteq \Sigma \ \& \ \alpha = \{l_1, l_2, ..., l_n\} \ \& \ [l_i = a_i \ or \ l_i = b_i] \ i = 1, 2, ..., n\}$$

Then we can think of each aggregate in T as a memory with n locations, where the ith location contains a 1 if $l_i = a_i$ and contains a 0 if $l_i = b_i$. We can 'execute' the following types of instructions: 'for all aggregates in the tube T, if location 3 has a 1 and location 5 has a 0, then set location 51 to 1' for example with the following molecular program:

(1) Input(T)

(2) $T_1 = +(T, a_3)$

(3) $T_2 = +(T_1, b_5)$

(4) $T_3 = T_2^{b_{51}}$

(5) $T' = \cup(-(T, a_3), -(T_1, b_5))$

(6) $T = \cup(T', T_3)$

3   The DNA computer

In this section we consider a hypothetical DNA based implementation of the 'restricted model' (see section 2) of molecular computer.

We will require that the alphabet of symbols Σ contain two special symbols $s_{5'}$ and $s_{3'}$. For each symbol $s \in \Sigma$ an oligonucleotide (i.e. short sequence of nucleotides, oligo) is chosen. The resulting set of oligos should satisfy the following properties:

(1) Under easily achievable conditions (for example of temperature, pH, salt concentration) each oligo reliably forms stable hybrids with its Watson-Crick complement.

(2) Under easily achievable conditions each oligo reliably dissociates from its Watson- Crick complement.

(3) Under neither of the conditions above does any oligo form hybrids with itself or another oligo, nor with another oligo's Watson-Crick complement.

For the DNA computer we will require that all input tubes consist of molecules of DNA which begin (5') with the oligo associated with $s_{5'}$ and end (3') with the oligo associated with $s_{3'}$.

If a *Merge* of tubes is required, this is accomplished by pouring the contents of all tubes into a single tube.

If a *Separate* is required on a tube T with respect to a symbol s, then if s is associated with the oligo O, a separator which consists of molecules of the oligo Watson-Crick complementary to O conjugated to solid supports are used. For example, the magnetic bead system employed in [1], or an affinity column.

If a *Detect* is required on a tube T, then PCR with primers appropriate for the common 5' and 3' sequences is performed followed by gel electrophoresis.

An Example Application:

To gauge what may be possible with a DNA computer, we will consider an example. Consider solving an instance of the 'satisfiability problem' [4]. Given a propositional formula ϕ with say 70 variables and 1000 binary connectives. Then using Lipton's method, $\Sigma = \{T_i | i = 1, 2, ..., 70\} \cup \{F_i | i = 1, 2, ..., 70\} \cup \{s_{5'}, s_{3'}\}$ - a 'true' symbol and a 'false' symbol for each of the variables and the required 'PCR symbols' $s_{5'}$ and $s_{3'}$. Associating each symbol of Σ with a randomly choosen (or carefully designed) 10-mer would seem a reasonable approach to satisfying the three conditions listed above. Denote the oligo associated with T_i as O_i^T, the oligo associated with F_i as O_i^F and the oligos associated with $s_{5'}$ and $s_{3'}$ as $O_{5'}$ and $O_{3'}$ respectively. One would want the input tube T to consist of the set of all DNA molecules of the form:

$$O_{5'} M_1 M_2 ... M_{70} O_{3'}$$

where for $i = 1, 2, ..., 70$, $M_i = O_i^T$ or $M_i = O_i^F$. Thus the elements of T are in one to one correspondence with the set of all possible truth assignments for the 70 variables.

---

[4]The 'satisfiability problem' is to distinguish satisfiable propositional formulas from those which are not satisfiable. For example '(A or B) and (not C and not A)' is a propositional formula with 3 variables (A,B,C) and three binary connectives (one 'or' and two 'and'). This propositional formula is satisfiable since the truth assignment which sets A to False, B to True, and C to False, makes the formula evaluate to True. The formula '(A and not A)' is not satisfiable since no truth assignment makes the formula evaluate to True. The 'satisfiability problem' is NP-complete and the best known algorithms for it apparently require an essentially exhaustive search of all possible truth assignments.

There would be 2^{70} molecules, each of which is a 720-mer. Assuming that each nucleotide has molecular weight 300 dalton, then the molecules in tube T have a mass of approximately 425 grams (0.93 lb.). While this is a rather large quantity by the standards of current molecular biology, there do not seem to be reasons in principle why dealing with such quantities would not be feasible in practice.

To create T one could proceed as described in [5] using the methods of [1] (see also [3]). Alternatively, it is possible to proceed in the following manner - well known in 'combinatorial chemistry' [4]. First, all of the O_i are produced separately. Initially, approximately 2^{70}, $O_{5'}$ are placed in tube T. Then the following steps are taken for $i = 1, 2, ..., 70$:

(1) pour half of the contents of tube T into T_1 and half into T_2. (2) To the molecules in T_1 ligate (a single) O_i^T to the 3' end. (3) To the molecules in T_2 ligate (a single) O_i^F to the 3' end. (4) Merge T_1 and T_2 into T.

Finally to the molecules in T ligate (a single) $O_{3'}$.

This is on a scale larger than customary in molecular biology and also creates 720- mers which are longer than currently produced by standard methods. Nonetheless, these problems appear technical in nature and creating T by this or other means seems feasible.

The required merges would be performed by 'pouring'.

Performing detection by PCR would be accomplished by standard means. Recall that such a PCR is only performed once at the end of the computation. If the final tube contained many DNA molecules (i.e. if there were many truth assignments which satisfied ϕ) then the detect operation could be performed with coarser techniques such as optical density measurement.

Performing separations might be done using the magnetic bead system of [1]. Each bead is approximately 1 micron in diameter and has a mass of approximately 1.3 pg. Each bead can bind approximately 7.8×10^5 biotinylated oligos. Assuming that one bead-associated biotinylated oligo will bind 1 target molecule, then to bind 2^{69} target molecules (the number of targets in the largest separation done during this computation) would require approximately 984g (approximately 2.2 lb.) of beads with a volume of roughly 0.39 liters (approximately 0.49 qt.)[5]. While separation by this method is on a scale unusual in

---

[5]Because spheres do not pack without spaces, a container for these beads would have a volume of approximately 750 ml (approximately 0.79 qt.)

molecular biology, it may be feasible. It is important to note that these
beads are reusable. After a separations, the beads and their oligos are
intact except for 'natural degradation'.

Assume that each separation could be done in 1000 sec. (i.e.
τ =approximately 16.7 minutes - this seems a plausible estimate) and
that the time for a merge is negligible. Also assume that there are no
errors during separations (i.e. $\epsilon_+ = \epsilon_- = 0$ - this is unreasonable, but
how non-zero values affect the results will be considered separately in
the following section (section 4)). Let an operation be taken as either
having a molecule of DNA 'decide' to go into one tube or another
during a separation, or having a molecule of DNA transferred from one
tube to another during a merge. The number of separations is 1001
and the number of merges is 1000. For a typical ϕ, it would seem
reasonable to assume that each separation and each merge would be
applied to a tube with approximately 2^{69} DNA molecules, hence the
number of operations per second would be approximately 1.2×10^{18}.
This is approximately $1,200,000$ times faster than the fastest super
computer. The total time for the calculation would be dominated by
the 1001 separations and would be approximately 11.6 days. During
the computation approximately 10^{24} operations would be performed,
which is possibly greater than the total number of operations performed
on man-made computing devices though out history[6].

It is also worth noting that while the molecular computer
above would require 140 'separators' (a 'True' and 'False' for each
propositional variable), Lipton's algorithm makes use of only one at a
time. Hence as many as 140 propositional formulas might be handled
by the machine simultaneously. In this case, the machine might execute
as many as 1.6×10^{20} operations per second.

4   Error Control

In this section we will consider the implications of errors during
separation. To begin, we will take care of a technical point. Each
incarnation of a restricted computer is associated with the parameters
$\tau, \epsilon_+, \epsilon_-$ (see section 2). It could be the case that ϵ_+ and ϵ_- are very
different. For example in the DNA computer described in section 3,

---

[6]We have not addressed the energy consumption aspects of molecular computers
in this paper. Nonetheless, one can expect that molecular computers will use much
less energy than electronic computers.

the process of separation might typically be carried out in two steps:

(1) *Anneal.* Incubate the contents of T with probes conjugated to beads and allow time for molecules with the correct consecutive subsequence to anneal. Pour off the liquid phase containing molecules that do not anneal.

(2) *Wash.* Add solvent and pour off. This will remove additional molecules which do not anneal (for example molecules left on the walls of the tube or held non-specifically to the beads). Repeat.

One can imagine that once a molecule that should be in $+(T, s)$ sticks it will stay stuck (with high probability) though many washings. So that ϵ_+ is determined by how well the molecules in $+(T, s)$ stick during the anneal step. On the other hand molecules which should be in $-(T, s)$ but which bind non-specifically may only be removed on a second or third washing. So ϵ_- may depend on how much washing is done. One can imagine a system with several washings which would give values like (these are strictly conjectural) $\epsilon_+ = 1/10$ and $\epsilon_- = 1/10^6$.

In the analysis that follows it will be preferable to have ϵ_+ and ϵ_- approximately equal. This is always possible to achieve (at least in theory) with the use of repeated separations:

Input(T) (1) $T_1 = +(T, s)$ and $T_1' = -(T, s)$. (2) $T_2 = +(T_1', s)$ and $T_2' = -(T_1', s)$ . . . (n) $T_n = +(T_{n-1}', s)$ and $T_n' = -(T_{n-1}', s)$ (n+1) $T^+ = \cup(T_1, T_2, ..., T_n)$ and $T^- = T_n'$.

Then after n steps T^+ will have $(1-\epsilon_+^n)$ of the elements which should be in $+(T, s)$ and will have $1 - (1 - \epsilon_-)^n$ of the elements which should be in $-(T, S)$. Further, $T = \cup(T^+, T^-)$. When n is choosen so that ϵ^n approximatly equals $1 - (1 - \epsilon_-)^n$ (which itself approximately equals $1 - n\epsilon_-$), then we can think of the n repeated separations, yielding the tubes T^+ and T^- from T, as a single new separation operation with ϵ_+ approximately equal to ϵ_- and τ n times as large as the original τ. For example, if the original separation had $\epsilon_+ = 1/10$, $\epsilon_- = 1/10^6$ and $\tau = 1000$ sec., then the new separation (with $n = 6$ - actually $n = 5.28$ is better but we will use an integral number of repetitions) has $\epsilon_+ = 1/10^6$, ϵ_- approximately $6/10^6$ and $\tau = 6000$ sec.

So hence forth we will assume that $\epsilon_- = \epsilon_+$ and we will denote this value as ϵ. We will assume that for our satisfiability example $\epsilon = 6/10^6$.

Now assume that a given molecule enters s separations in the course of a computation (e.g. in the satisfiability example $s \leq 1000$). What is the probability p_{good} that this molecule goes through all separations without a mistake (i.e. without ever going into the 'wrong' tube). That probability is:

$$(1 - \epsilon)^s$$

and the probability p_{bad} that it fails to go through the s separations without a mistake is:

$$1 - (1 - \epsilon)^s$$

For the satisfiability example, for all molecules $p_{good} \geq (1 - 6/10^6)^{1000}$ which is approximately 0.994 and $p_{bad} \leq 0.006$. Hence at least 994 molecules out of each 1000 make it through the entire calculation with no error. This is pretty good, but not adequate, because $\frac{6}{1000}2^{70}$ (approximately 1.5×2^{62}) molecules could have errors at some point and might end up in the wrong tube when the computation is done. To handle this problem consider the following:

Assume you have a traditional sequential electronic computer. There is a predicate P (i.e. a function which on each input returns either 'yes' or 'no') and assume you write a program which exhaustively (and sequentially) searches all 2^{70} binary strings of length 70 for the existence of a 'winner' string for which P gives a 'yes' answer. Further assume that for each string of length 70, P can be computed with 1000 operations. If during the search, the computer stops and prints out 'α is a winner', should you believe it? Like molecular computers, electronic computers are physical systems which are subject to error. If you are told that your computer makes at most one error in $2^{1,000,000}$ operations, then, as a practical matter, you are likely to be justified in believing that α is indeed a winner. If however, you are told that your computer makes at most one error every 166,666 operations, then the situation is not so clear. You may want to check the computer's answer by retesting the string α several times. Each time you get the answer 'α is a winner' (without ever getting the answer 'α is a loser'), you are justified in increasing your belief that α is indeed a winner - but of course you will never be absolutely sure.

These ideas will now be applied to the molecular setting. Assume that p_{good} is fairly large (as in the satisfiability example) and that we have run the computation for the propositional formula ϕ and have produced the output tube T_1^w which is supposed to (i.e. if there were no errors) have the 'winners' (i.e. those molecules which encode satisfying truth assignments for ϕ). What if we now take the tube T_1^w and use it as the input to an entire rerun of the whole computation (i.e. retest all of the purported 'winners')? We will then get a new tube T_2^w which is all of the molecules which went through the entire computation twice and both times ended in the winning tube. What if we repeat that process through r runs of the entire computation - what now will be in T_r^w? It will contain two types of molecules: the 'true winners' (those that really do encode satisfying truth assignments) and the 'false winners' (those which encode unsatisfying truth assignments - but have somehow, through errors, made it into T_r^w anyway). Notice that it is much easier to get into T_r^w as a 'true winner' than as a 'false winner'. A 'true winner' need only go through r computations without an error - but most molecules do make it though each pass of the computation without an error (in our example 994 of every 1000 at least). In fact, the probability of making it through r passes is at least p_{good}^r. On the other hand, a 'false winner' is unlikely to make it through to T_r^w. A 'false winner' must have at least one error on each pass through the computation (otherwise on the pass without errors, it would end where it belongs - out of the winner tube). But only at most p_{bad} (in our example 6 in 1000) make an error in each pass. To make an error r times in a row has a probability of at most p_{bad}^r. In our example we could choose $r = 10$. Then p_{bad}^{10} is approximately $1/(13.5 * 2^{70})$. And hence the expected number of 'false winners' in tube T_{10}^w is at most $1/13.5$ (less than one). Hence even if ϕ is not satisfiable, it is unlikely some 'false winner' will end up in T_{10}^w and make us believe that it is satisfiable (of course we could always detect a 'false winner' anyway, by checking if the truth assignment it encodes really does satisfy the formula). Now p_{good}^{10} is approximately 0.94. So even if there is just one satisfying truth assignment for ϕ, there is at least a 94% chance the molecule encoding that assignment will end up in T_{10}^w - and hence we will be likely to give the correct answer, that ϕ is satisfiable.

It is important to note that the analysis above was rather heavy handed and that a more refined analysis would probably give a better value for r. In addition, there may well be better ways to handle separation errors than those proposed above. It seems unlikely that the approach described is optimal either mathematically or physically. On the physical side there is probably much room for improvement. The sorts of separations actually needed are quite special. First, the oligos used can be specifically choosen to diminish the possibility of non-specific binding. Second, the tubes are free of extraneous molecules (e.g. enzymes) which might complicate the interactions. Third, there is a symmetry in this setting. For example, if a separation with respect to the symbol F_5 is done then $-(T, F_5) = +(T, T_5)$, since every molecule in the system has either T_5 or F_5 but not both. Consequently one could imagine carrying out the separation using a 'double positive' system. Consider putting the contents of the tube into a chamber which contains beads for T_5 on the left and beads for F_5 on the right with a semipermeable membrane in between which allows oligos to pass but not beads (indeed with such a system, the beads might be replaced by large molecules which cannot pass the membrane). Agitation of the chamber might yield very clean separations - experimentation would allow one to be sure. Such a system might also be very fast and perform separations without the need to expand the volume of solvent used. At any rate, even the analysis done here gives reason to be encouraged that separation error control will not be an onerous burden.

There are of course other possible sources of error in molecular computation; for example, 'natural degeneration' of molecules. In a DNA computer, it would of course be necessary that major sources of degeneration such as nuclease contamination be avoided and it would be sensible to have separations done under the mildest conditions possible. It might also be wise to use analogues of DNA such as those with peptide backbones which might be more robust. In any case, it seems unlikely that a reasonable system for controlling such errors could not be designed.

5　Other Incarnations: The Non-biological Catalytic Molecular Computer

In this section the prospects for an improved incarnation of a molecular computer are considered.

Because the amplify instruction is not used in the restricted model, there may no longer be a need to use biomolecules for computation. Using non-biological molecules may have several advantages.

First, these molecules could potentially be much smaller than those used in the DNA computer. In the DNA computer each symbol is associated with a 10-mer which has a molecular weight of approximately 3000 dalton. In a non-biological computer one might hope to associate each symbol with a much smaller molecule. Since roughly speaking mass is inversely proportional to the amount of parallelism available, such small molecules could lead to restricted molecular computers with much greater parallelism than the DNA computer.

Second non-biological molecules might be constructed which were extremely stable. This could result in a 'catalytic' molecular computer. Since merge and separate do not involve the making or breaking of covalent bonds, if we assume that detect can be done without the construction or destruction of molecules, then once a computation is completed, all resulting tubes need only be merged into a single tube to recreate the input. Hence the molecular computer in a real sense catalyzes the computation. This may be important as a practical matter. For example, one could build a molecular computer for satisfiability which would create the input tube T (consisting of the molecules encoding all possible truth assignments) just once. Then given a formula ϕ, T would be used to determine whether ϕ was satisfiable. Once that computation was completed, the tube T would be recreated and the next formula processed[7].

Third, given the freedom to choose arbitrary molecules, it may also be possible that a set of non-biological molecules might be choosen with properties that allow for systems with small values of τ, ϵ_+, and ϵ_-.

What an appropriate set of molecules would look like is best left to professional chemists. However, to provide an unsophisticated starting point for discussion, consider the following system. Associate each symbol with a small functional group (e.g. methyl, amino). Form an aggregate by bonding the appropriate functional groups to a fullerene skeleton. For the purposes of separation, use 'combinatorial' chemistry or rational design to develop enzymes which specifically bind to each incorporated functional group. Such a system, if possible, would encode an aggregate in approximately one fiftieth of the mass used by the DNA computer. Hence it would provide approximately 50 times the parallelism. Further fullerenes are apparently quite stable.

The memory model might (in theory) be implemented by a system like the following. For each 'location' i, a_i is associated with an oligonucleotide O_i and b_i is associated with a methylated version of O_i. Enzymes are used to methylate and demethylate. For example, if $O_i = 5'...GAATTC...3'$ then EcoR I methylase will catalyze the

---

[7]By implementing *Detect* carefully on a DNA computer, it seems that it might be made 'catalytic' or nearly so.

transfer of a methyl group (from S- adenosylmethionine) resulting in $5...GA^mATTC...3'$. EcoR I methylase will not methylate oligos without the subsequence $GAATTC$. Unfortunately, specific enzymes to demethylate appear not to be known.

A bio-molecular incarnation, might also be possible. As above, each a_i is associated with some small functional group and b_i is obtained from a_i by the non-covalent attachment of a small molecule which binds to a_i.

Building a computer with a small τ (i.e. time of separation, see section 2) would be very desirable. One of the problems with the designs described to this point is that 'separation' is a largely mechanical process applied from outside the tubes. It would be interesting to design a molecular computer (perhaps based on entirely different 'primitives' than those used here) which would accomplish its task by purely chemical means inside of a single tube.

Since the restricted model of computation requires only merges, separations and detections, there may be other physical systems which could provide an incarnation. For example, some form of all of these operations appear possible using atomic or subatomic particles traveling in accelerators with separation done according to mass or charge.

6   Detect

In going from an unrestricted computer to a restricted one, the amplify operation was removed. Nonetheless in the DNA computer, amplification was still used in order to implement the detect operation. If a restricted computer based on some other material is to be built, then some means of performing detect without amplification may be required. There appear to be reasons to be optimistic about the existence of such systems. It may be possible to detect single molecules in solution using fluorescence labeling[5]. It appears that physicists are able to detect a single subatomic event in a huge volume of space when detecting neutrinos.

Another operation which might be considered in a molecular computer is *Describe*. Roughly, decode a molecule back into the set of symbols it encodes. For example, when determining whether a propositional formula ϕ is satisfiable or not, a molecular computer may produce a tube which contains the molecules that encode satisfying truth assignments (if such exist). One may wish to have an explicit description of one such truth assignment (rather than to just know that at least one exists). Hence one would want to take a molecule from the tube and *Describe* it. This operation would be implementable in the DNA computer by well know methods (e.g. sequencing) of molecular biology. It is worth noting however that it is in fact efficiently implementable in all incarnations of restricted molecular computers. Let the alphabet of the system be $\{s_1, ..., s_z\}$, then to *Describe* a single molecule which is contained in a tube T, one can use the following program:

(1) Input(T)
(2) Symbols= $\emptyset$ (begin with the empty set of symbols)
(3) If Detect(T) = 0 then output Symbol.
(4) For $i = 1$ to z:
    (a) If Detect($+(T, s_i)$) = 1 then:
        (i) $T = +(T, s_i)$
        (ii) Symbol=Symbol$\cup\{s_i\}$ (add an s_i to the set
                of symbols)
    (b) If Detect($+(T, s_i)$) $\neq$ 1 then:
        (i) $T = -(T, s_i)$
(5) Output Symbol

Notice that even if T contained many different molecules each encoding a different set of symbols, the program above would give an explicit description of exactly one of them.

7 Applications to Biology, Chemistry and Medicine

It is the author's hope that the ideas used in designing and programming molecular computers will be of direct (non-computational) value in biology, chemistry and medicine. The ability to create molecules with desired properties (e.g. enzymes, drugs) 'on demand' is of great importance in these areas. Historically, the process of creating such molecules has been difficult (if possible at all) and expensive. Very recently, a new promising approach to this problem, called 'combinatorial chemistry' [4], has arisen. Below is an example of its use. It should be clear that combinatorial chemistry is closely related

to molecular computation. It seems possible that concepts useful
in designing molecular computers may be of value in combinatorial
chemistry and *vice versa*. Following the example a potential application
of a computational concepts is described.

Recently, Bartel and Szostak [3] have used the methods of
combinatorial chemistry to make what might be called a 'pseudo-
enzyme'. We will describe their basic approach briefly. For the sake of
clarity many simplifications will be made and many important details
will be omitted.

The goal of the experiment was to find a molecule of RNA which
would ligate two substrate molecules of RNA ρ_1 and ρ_2 (ρ_1 and ρ_2
were base pair complementary in such a way that hybridization would
bring their 5'- triphosphate and 3'-hydroxyl groups into proximity
in preparation for ligation). A pool of approximately 4^{25} random
sequences of RNA was developed. Each RNA in the pool was bound
to a copy of ρ_2 at the 5' end and a copy of a constant region C at
the 3' end. Hence, a tube containing approximately 4^{25} molecules was
created, and each molecule in the tube had the form: $\rho_2 RC$ where
R was some molecule of RNA from the original pool. To this tube
an excess of ρ_1 was added. If a 'winner' W existed in the RNA pool
which could ligate ρ_1 to ρ_2 then in the tube $\rho_1 \rho_2 WC$ would be formed.
Next molecules which contained the sequence ρ_1 were separated from
the rest and retained (hence 'non winner' RNA from the original pool
were removed). Because the 5' sequence of ρ_1 was known and the
sequence C was known, it was now possible to amplify (using reverse
transcription, PCR, transcription) only those molecules of the form
$\rho_1 \rho_2 WC$. Standard means could now be used to discover the sequences
of the 'winners'.

The 'winner' of Bartel and Szostak experiment is not a true enzyme,
since it is 'used up' in acting on its substrates. The value of having
the 'winner' RNA 'anchored' to one of the substrates is that once the
winner acts, it becomes permanently anchored to the reaction product.
This distinguished it from the rest of the RNA pool and allows it to be
retrieved for later identification. If one wished to create a true enzyme,
one might have to identify the 'winner' despite the fact that after acting
on its substrate it reenters the pool. This is an example of the problem
which we will now explore further: finding the 'winner' when all that
is available is the ability to know that a 'winner' exists somewhere in a
large pool. Being able to solve this problem in a general setting would
be particularly important when the pool molecules were not nucleotides

(e.g. a pool of proteins, a pool consisting of multiple variants of an existing compound). Or when anchoring was not physically possible or appropriate (for example when an endonuclease was sought or a drug which crosses a cell membrane and then binds a particular host molecule). Below we show that this problem can be solved in a wide variety of settings. We demonstrate the technique with an example. Our goal will be to find an RNA molecule which will ligate (as a true enzyme) two DNA molecules δ_1 and δ_2.

We will need a little notation. Let P denote the pool of all 25-mers of RNA (there would be a total of 4^{25}). If we have an RNA sequence σ let P_σ be all 25-mer RNA which begin (5') with σ. For example P_A would be all 25-mers which begin with A (there would be a total of 4^{24}), and P_{ACCU} would be all 25-mers which begin with $ACCU$ (there would be a total of 4^{21}).

One can begin as in [3] by creating an RNA pool consisting of all 25-mers (i.e. begin with P). One then incubates P, and excess δ_1 and δ_2 in an appropriate buffer. After an appropriate time, one runs a PCR with primers appropriate for the 5' end of δ_1 and the 3' end of δ_2. Thus neither δ_1 nor δ_2 will be amplified but the product of their ligation would be. Hence if no PCR product is produced we will declare that the pool P contains no 'winner' and discontinue the experiment for that pool. If a PCR product is produced we will declare that the pool P contained a 'winner'[8] How do we find the winner? We proceed as follows. In the next step we repeat the experiment above 4 times, once with each of P_A, P_C, P_G and P_U. Since $P = P_A \cup P_C \cup P_G \cup P_U$, at least one (and perhaps more than one) of them will contain a 'winner'. Lets say P_C contains a 'winner', then we know some 'winner' begins with C. Next we repeat the experiment with P_{CA}, P_{CC}, P_{CG} and P_{CU}. Again at least one must contain a 'winner'. Say P_{CA}, then we know some winner begins with CA. Continuing in this fashion we will come to know the exact sequence of some 'winner' as desired.[9]

Admittedly this is a considerable amount of repetitious work. It requires 101 steps of pool synthesis, incubation and PCR (many of these steps could be run in parallel of course). However, in these 101

---

[8]For the purpose of this illustration, we will ignore certain 'practical' problems such as the occasional uncatalyzed ligation of δ_1 and δ_2.

[9]The complexity theorist will recognize the similarity of this technique to that used for demonstrating that if there exists a polynomial time algorithm for deciding satisfiability, then there exists a polynomial time algorithm for finding satisfying truth assignments.

steps, 4^{25} (over 1000 trillion) RNA strings are searched. This is of course the same sort of trade off achieved for our molecular computers applied to problems like satisfiability (linear time versus exponential time). Further, by doing the incubation under 'unfavorable' conditions where only the 'best' winners will have a chance to act, the need for 'evolving' the answer as done for example in [3] might be removed and this should save considerable time. Also, there is never a need to perform amplification on the elements of the pool (though in this example we do use amplification on the product of the ligation to detected if a winner exists), which in the case of RNA is laborious and for other molecules may be impossible.

It is worth remarking that this technique would work in the same way if the pool was composed of RNAs consisting of all length 25 sequences of elements from any set of four 'basic' RNA units (for example four 10-mers rather than the 1-mers used in the example). The RNA in the pool could also have fixed constant regions if that was desired.

It should be clear that similar techniques would work in a wide variety of settings. Using pools made of RNA, DNA, proteins or some other form of aggregates. Whether a ligation or some other function was desirable - so long as the existence of a 'winner' could be detected.

So the method demonstrates that the ability to detect the existence of a winner leads efficiently to the identification of a winner.

8   Discussion

First a general caveat. To correctly gauge the practicality of molecular computation will require inputs from experts in a wide variety of fields including: biology, chemistry, computer science, engineering, mathematics and physics. The author is not well versed in all of these areas, yet has choosen, either explicitly or implicitly, to make assumptions about each in writing this paper. Errors and omissions should be expected. Ideally, those with expertise in these areas will find ways to improve the approach or reveal fundamental flaws.

Can a practical molecular computer be built? It is still too early to know. However, the DNA computer described here gives some reason for optimism.

The author suspects that molecular computers may fill a computational niche which electronic computers do not and vice versa. Is it better to have one computer which can execute a trillion instructions per second, or three computers which can each execute half a trillion instructions per second? The answer depends of what problem you wish to solve. If the problem (e.g. factoring an integer) can be split into three tasks each requiring about the same number of operations to complete, then the three computers will solve the problem more quickly. If however the task is not decomposable (e.g. computing the trillionth digit of ϕ by existing algorithms) in this way, then the one faster computer is best. Molecular computing takes this example to an extreme, instead of three computers executing half a trillion instructions per second, you are offered 10^{20} (or more) computers each of which executes one instruction every 1000 seconds or so. Which is best? If your problem can be decomposable into a huge number of tasks and each task is very simple (as in our example of satisfiability) then the molecular computer may be much faster. If such a decomposition is not possible then the single faster computer is best. It is an interesting challenge to envision a system somewhere in between, which provides a moderate number of computers each operating at a moderate speed.

9 Acknowledgments

The author would like to thank Jonathan DeMarrais, David Jefferson and Ron Rivest for their comments.

Bibliography

[1] Adleman L. Molecular computation of solutions to combinatorial problems. *Science* 266:1021-1024 (Nov. 11) 1994.

[2] Lipton R. Speeding up computations via molecular biology. Draft. Dec. 9, 1994. (by anonymous ftp: /ftp/pub/people/rjl/bio.ps on ftp.cs.princeton.edu).

[3] Bartel D. and Szostak J. Isolation of new ribozymes from a large pool of random sequences. *Science* 261:1411-1418 (Sept. 10) 1991.

[4] Alper J. Drug discovery on the assembly line (Research News).*Science* 264:1399-1401 (Jun. 3) 1994.

[5] Eigen M and Rigler R. Sorting single molecules - applications to diagnostic and evolutionary biotechnology. *PNAS* 91: 5740-5747 (Jun. 21) 1994.

Eric B. Baum

DIMACS Series in Discrete Mathematics
and Theoretical Computer Science
Volume **27**, 1996

A DNA Associative Memory
Potentially Larger than the Brain

Eric B. Baum
NEC Research Institute
4 Independence Way
Princeton, NJ 08540

`baum@research.nj.nec.com`

Abstract

We propose a method for making a large content addressable memory using DNA and discuss the prospects and potential of such techniques. An associative memory vastly larger than the brain may be possible.

Leonard Adleman's seminal paper[1] has sparked a number of proposals for using DNA to solve computational problems. Several are described in this volume. This paper will emphasize a different application- using DNA to build a large associative or content addressable memory. This proposal was also described in [2].

I consider a memory content addressable if a stored word can be retrieved from sufficient partial knowledge of its content, and associative if a stored word can be retrieved from an input cue which may in fact contain errors. So for example, if we stored the three words

$$0, 0, 0, 0, 0, 0, 0, 0, 0, 0$$
$$1, 1, 0, 1, 0, 0, 1, 0, 1, 1$$
$$0, 0, 0, 1, 1, 0, 0, 1, 0, 0$$

in a content addressable memory, and then presented an input cue

$$*, 1, *, 1, *, *, *, *, *, *$$

we would hope to retrieve the second stored word, and then be able to read out its entire content. Here * means "don't care"- so this input cue has information only about the second and fourth components. In an associative memory, you might be presented with an input cue

$$*, 1, 1, 1, *, *, *, *, *, *$$

which doesn't match any of the stored vectors exactly, and would hope still to retrieve the second stored word as the closest match. Standard computer memories are neither content addressable nor associative-instead you need to know a specific address to retrieve a stored word.

The brain's memory seems to be highly associative. Thus most readers, given the cue "the big, gruff actor who starred in the movie 'The Green Berets' " have no trouble recalling everything they know about John Wayne, including possibly images of him, scenes from movies, and anecdotes, and they could equally well have recalled everything they know about him from many, many other cues, including those containing some errors. Associative memory is frequently thought to be, in fact, a key factor in human intelligence; hence there has been a large literature in the neural net community on associative memories. Hopfield's paper[4] on neural net associative memories had a large sociological impact in the rennaissance in neural networks in the '80s. A paper of myself, Moody, and Wilczek[5] presented and analyzed some efficient neural designs for associative memories. The present paper will describe related designs using DNA instead of electronic circuits.

Building a content addressable memory using DNA is conceptually straightforward. The memory consists of a flask. You store words of a fixed length. To write such a word you place in the flask a DNA molecule encoding it. One may for this purpose utilize an encoding similar to that proposed by Lipton [3] for Satisfiability. In such an encoding, there is a specific DNA subsequence assigned to each component, value pair. To store a word you append the DNA subsequences corresponding to each of its components together to form a molecule. These subsequences can have been previously mass produced, say by PCR, and so producing this molecule will be inexpensive. When the molecule is placed in the flask, the word is written.

To retrieve a word given a cue, you retrieve its associated molecule. This can be done content addressably by a sequence of extract steps. For each component in the input cue, you introduce the complementary subsequence affixed to magnetic beads. These sequences then stick to molecules encoding words containing that component value. These molecules can then be extracted magnetically. After sequential extractions for each of the component values in the cue, one is left with molecules matching the cue. The bead extraction technique was used by Adleman[1] to extract molecules coding for Hamiltonian Paths. Once the molecule is retrieved, it can be sequenced to read the word in its entirety.

To achieve an associative (as opposed to merely content addressable) memory it would be necessary to find a better marking procedure than using magnetic beads. Ideally one would like to simultaneously introduce into the flask subsequences complementary to all the sequences in the cue, each marked somehow. Then the molecule in the memory with the most marked sequences sticking to it is the closest match to the cue. This would also make for a faster recall than if the retrieval must be done component by component. This would be logically analagous to the grandmother memory of [5]. Unfortunately the biotin beads are too large- you can't have many of them sticking to one DNA molecule at the same time. But an alternative marking technology may prove possible.

A number of refinements may be suggested. It may be preferable to only use subsequences coding for '1's, and not bother with subsequences coding for '0's. In this case, a stored word would correspond to a molecule in which the appropriate subsequences for its '1's were appended together. This would save DNA, and make it possible to store very long, sparse vectors, i.e. vectors with few '1's and many '0's. Such vectors may correspond well to human memories- we do not want to store in memory all the attributes that a particular object does not have, such as the fact that this book is not liquid. Similarly this feature may be important in data base applications.

There is considerable freedom in choosing the subsequences. This freedom may be employed to choose subsequences which facilitate retrieval and reading. Reading does not require full sequencing of the DNA molecules, but rather only involves deciding which component subsequences are present. Thus we may wish to choose subsequences to be readily distinguishable, say chosen from an error correcting code. This would also make less likely accidental binding of complementary subsequences during the retrieve step, which may be a significant problem. Also the subsequences may be chosen to facilitate specific reading techniques such as restriction mapping[6]. For more details see my article in Science[2].

We may choose to store molecules only part of which are content addressable. Appended to these might be sequences encoding more compactly coded information- perhaps even written directly in base 4. In the limit of this, we might in fact simply use DNA to form a standard random access memory. In such, each word would have an address portion and a data portion. The data portion could be both quite long and compactly coded. To retrieve the word we introduce the subsequence exactly complementary to its address portion. The address portion could also be rather longer than individual component subsequences would likely be, and if the addresses were taken from an error correcting code misretrievals would be quite unlikely. The data portion might be kept double stranded, and the address portion single stranded. This may speed up annealing- i.e. retrieval, and would alleviate problems such as hydrolysis(i.e. the fact that single strand DNA molecules in water for long periods of time will dissolve and break.) If problems occur in double strand DNA, they can in principle be corrected by "proofreading" techniques.

In principle humongous capacities may be achieved by DNA memories. A standard concentration of DNA in water is .06g/liter[7]. A millimole of 200 base long molecules would weigh about 50g and thus occupy about 1000 liters, roughly a large bath tub. If you can really get away with one molecule per stored word, this would mean a storage of well over 10^{20} words! Even if it is neccessary, because of practical considerations mentioned shortly, to use many thousands of copies of each molecule, you would still have an immense memory. By comparison, the brain has perhaps 10^{15} synapses, and so by conventional estimates probably stores not much more than about 10^{15} bytes (although we should not neglect the possibility that the brain encodes information at a molecular level either). I have heard Feynman credited with the estimate that the brain distinguishes about 10^{6} concepts. This comes from the observation that 20 questions is an interesting game- we play 20 questions, not, say, 30 questions. Using 20 yes-no questions optimally, it is only possible to distinguish $2^{20} \sim 10^{6}$ different concepts. This is an intriguing estimate, indicating perhaps that our minds are less complex than we usually think. It serves anyway as a lower bound on the number of concepts stored in the brain.

The DNA Associative memory would have very slow retrieval times, at least using present technology. Standard biological operations such as extraction take, currently, hours to perform.

It is not immediately evident what the best applications for a humongous, very slow, associative memory are. Probably there are interesting applications to retrieval on huge databases.

Will this be a reality? Surely it could be built at a small scale. The question is, can we overcome the hurdles to scaling it up to interesting size? These are the same as afflict all the DNA computing approaches to date, as have been emphasized particularly by Smith and Schweitzer.

See e.g.[8],[9] for discussion. Will annealing times (i.e. the time it takes a subsequence to find its complementary subsequence) slow down unacceptably as solutions are scaled up? This possibility will likely force, at least, use of multiple copies of each word. Will sticking of sequences to approximate matches, as opposed to exact matches, glom up the system too much? Will the DNA hydrolyze too fast? To make DNA computing a reality, these problems will have to be addressed. Life has addressed such problems in vivo, so they would seem to be solvable at least in principle. Biologists have come up with some tricks that life has presumably not used[1]. It remains to be seen whether DNA computing or DNA Associative memories will be a reality or a footnote.

Acknowledgement: I thank Peter Kaplan, Warren D. Smith, and Allan Schweitzer for helpful comments.

Bibliography

[1] Adleman, L. M., "Molecular computation of solutions to combinatorial problems", Science Vol 266, 1021-1024 (1994).

[2] Baum, E. B., "Building An Associative Memory Vastly Larger Than the Brain", Science V268,583-585 (1995).

[3] Lipton, R. J., "DNA Solution of Hard Computational Problems", Science V268, 542-545 (1995).

[4] Hopfield, J. J., "Neural Networks and Physical Systems with Emergent Computational Properties", PNAS USA 79:2554.

[5] Baum, E. B., J. Moody and F. Wilczek, "Internal Representations for Associative Memory", Biological Cybernetics, v59 (1988), 217-228.

[6] Alberts, B., D. Bray, J. Lewis, M. Raff, K. Roberts, and J. D. Watson, *Molecular Biology of the Cell*, Garland Publishing, New York, 1994.

---

[1]For instance, apparently T. J. Meade and J. F. Kayyem have a proposal for identifying DNA using its electric conductivity properties,[10]

[7] Sambrook, T., E. F. Fritsch, T. Maniatis, *Molecular Cloning, A Laboratory Manual, 2nd Edition*, Cold Spring Harbor Press, Plainview NY, 1989.

[8] Smith, W. D., A. Schweitzer, "DNA Computers in Vitro and Vivo", to be published.

[9] Smith, W. D., "An opinionated, but reasonably short, summary of the Mini DIMACS Workshop on DNA based computers, held at Princeton University on April 4, 1995." Unpublished manuscript.

[10] Paterson, D., "Electric Genes, current flow in DNA could lead to faster genetic testing", Science and the Citizen, Scientific American May 1995.

DIMACS Series in Discrete Mathematics
and Theoretical Computer Science
Volume **27**, 1996

A Universal Molecular Computer
(condensed abstract for DIMACS Workshop of April 4, 1995)

Donald Beaver[1]
Penn State University
317 Pond Lab
Penn State University
University Park, PA 16802
(814) 863-0147
beaver@cse.psu.edu.

Abstract

We design a molecular Turing machine and determine the complexity of the problems solvable by molecular computers.

Interest in "nanocomputation" has been sparked by Adleman's recent experiment demonstrating the possibility that molecular computers might solve intractable problems, such as Hamiltonian Path, using large-scale parallelism achievable only through molecular-scale miniaturization.

We propose a method for site-directed mutagenesis (namely, a molecular "editing" reaction) and use it to build a universal computer, stepping beyond Adleman's special-purpose, one-time problem solver. Using the generous assumptions on parallelism implicit in Adleman's methods, we show that molecular computers can in fact compute PSPACE. Under stronger and more realistic restrictions, we show that molecular computers – both ours and Adleman's – are limited to solving problems in BPP.

1   Introduction

In the same issue of *Science* in which Adleman presented his results on solving Hamiltonian Path using DNA [1], Gifford posed the question of whether general-purpose computers based on DNA are possible [8]. In November, 1994, we then immediately developed a method to solve NP and moreover to build a general-purpose computer, at least via a feasible thought-experiment.

---

[1]Research supported by NSF CCR-9210954

The tape of a Turing machine bears strong similarity to DNA strands: it is linear and stores information over a finite alphabet, $\Sigma_{dna} = \{A, C, G, T\}$. We represent the configuration (tape and current state) of a TM as a linear, double-stranded sequence, like a chromosome.

Each new generation is derived from its parent through a mutation that implements one step of a Turing machine. The new generation represents one evolutionary step toward the final result. Interestingly, our complexity-theoretic characterization of these methods can be regarded in a broader sense as a form of sexual reproduction, in which portions of chromosomes are exchanged and only certain combinations survive.

While single strands of DNA may seem more appropriate, they are less stable, although Smith and Schweitzer do provide an elegant solution based on single strands [15]. Using circular strands permits other nice tricks and adaptations, all fairly easy variations (in thought-experiment form) for the researcher versed in recombinant DNA and complexity theory (*cf.* [12]). Our results were intended to demonstrate the possibility of DNA-based universal computing, thus we have chosen to describe our results in the more natural and appealing form of "chromosomes."

2   Turing Machines

A Turing Machine (TM) consists of a finite set Q of states, a finite alphabet Σ, a transition function δ mapping $Q \times \Sigma$ to $Q \times \Sigma \times \{L, R\}$ (where L and R denote left- and right-moves, respectively), and an infinite worktape (*i.e.* an infinite sequence $\sigma_1, \sigma_2, \ldots$ of symbols over Σ). We refer the reader to [9] for details.

For simplicity, we consider a $S(n)$ space-bounded TM, where n is the size of its input. The elements of Σ and Q can be encoded in a position-dependent form using $O(\log n)$ nucleotides in such a way that unintended overlaps do not occur. In particular, let $E(a, i)$ be the encoding of element a at position i, and let $\bar{E}(a, i)$ be its Watson-Crick complement. To ensure that only intended pairings occur, $E(a, i)$ should not overlap by more than 10 percent with $E(b, j)$ or $\bar{E}(b, j)$ for all b, j (excluding $\bar{E}(a, i)$, of course).

The current state and tape contents of a TM constitute a *configuration*. If the location of the tape head is i, the current state is q, and all symbols after location $S = S(n)$ are 0, we may write the configuration in a standard way as $\sigma_1 \cdots \sigma_{i-1} q \sigma_i \sigma_{i+1} \cdots \sigma_S$, and encode it using DNA as

$$E \ (\sigma_1 \cdots \sigma_{i-1} q \sigma_i \sigma_{i+1} \cdots \sigma_S) = E(\sigma_1, 1) \cdots E(\sigma_{i-1}, i-1) E(q, i)$$
$$E \ (\sigma_i, i) E(\sigma_{i+1}, i+1) \cdots E(\sigma_S, S).$$

Our goal, to implement a universal computer using DNA, reduces to editing the configuration according to the local computation specified by δ.

In particular, if C is a configuration, let δC be the configuration derived by executing one step of the TM. For example, if $\delta(q, 1) = (r, 0, R)$, then

$$\delta(0 \cdots 0 q 11 \cdots 1) = 0 \cdots 0 0 r 1 \cdots 1),$$

noting that there are $i + 1$ symbols to the left of r. Our goal is to construct a new molecule having sequence $E(\delta C)$ from an old molecule having sequence $E(C)$.

While it would be possible to read the entire molecule and simply build a new one with the appropriate modified sequence, this would merely be a complicated way to implement what electronic machines already trivially implement, at tiny cost and great accuracy. Instead, we read only the local configuration (near the tape head) and invoke a direct editing reaction. The advantage is that the rest of the configuration is unknown – and unrestricted. Thus, a host of configurations, each having the same local description (state q reading σ_i in position i) can be advanced simultaneously, even though the rest of their tapes are very different. This supports a degree of parallelism unheard of in electronic computation and provides a powerful return for our cumbersome experimental steps.

Site Directed Mutagenesis

We implement a specific form of site-directed mutagenesis. The context of the substitution (resp. deletion, insertion) should be known. That is, we seek to replace $\alpha X \beta$ by $\alpha Y \beta$, where α and β are substrings that do not occur elsewhere in the strand.

Roughly speaking, we perform a few "simple" steps. First, the original "chromosomes" are denatured and the single strands

containing $\alpha X\beta$ are retained. These strands are mixed with $\bar{\alpha}\bar{Y}\bar{\beta}$, ·
forming duplexes in which the α portions line up, the β portions line
up, but the $\bar{Y}$ and X portions remain open and unaligned. Then,
through PCR (the polymerase chain reaction), the remainder of the
original strand is copied.

We now have double strands which are aligned properly everywhere
except for the region between α and β, where X and $\bar{Y}$ occur. We
again denature these strands, except this time we retain only the newly
formed strands containing $\bar{\alpha}\bar{Y}\bar{\beta}$. Finally, these strands are copied
(and amplified, if needed) into double-stranded "chromosomes" again.
The $\bar{\alpha}\bar{Y}\bar{\beta}$ region gives rise to a $\alpha Y\beta$ region, rather than the original
$\alpha X\beta$ region. Everywhere else, the molecule(s) are identical to their
progenitors. (See [3, 5, 6] for details.)

If the $\bar{\alpha}\bar{Y}\bar{\beta}$ segments are small, then alignments will occur preferably
as desired. It remains possible that the $\bar{\alpha}$ aligns with one progenitor
while the $\bar{\beta}$ aligns with another, forming unintended aggregates. One
way to avoid this is to maintain a distance between progenitors by
binding them temporarily and non-covalently to single strands attached
covalently to beads or to a DNA chip. Smith [personal communication]
has pointed out that his solution (attaching DNA permanently to
a surface) achieves the same ends [15]. Another way is to use
circular strands, in which case the undesired alignments give rise to
molecules having improper length. These undesired products can then
be separated using gel electrophoresis.

Nondeterministic and Parallel TM's

Assume we have a test tube containing a set $T = \{C\}$ of possibly-
different configurations C of the same TM. For each state q, head
position i, and symbols $\sigma_{-2}, \sigma_{-1}, \sigma, \sigma_{+1}, \sigma_{+2} \in \Sigma$, extract those
configurations matching this combination and place them in a separate
tube, $T_{q,i,\sigma_{-2},\sigma_{-1},\sigma,\sigma_{+1},\sigma_{+2}}$. For each such tube, choose α, X, Y, and
β according to $\delta(q,\sigma)$, and invoke the appropriate editing reaction.
Finally, mix the results together. We obtain a tube δT containing
$\{\delta C | C \in T\}$.

Since there are $O(S(n))$ different tubes, there are $O(S(n))$
steps. (Examining all five neighboring symbols is clearly unnecessary.
Obtaining $O(\log S(n))$ steps through parallelization is a simple exercise.
Many refinements have been left to the full version.)

Thus, a universal computer can be implemented with $O(1)$

overhead. More importantly, a heterogeneous set of configurations can be advanced simultaneously and in parallel. One way to model a nondeterministic TM is as a deterministic TM assisted by an extra, nondeterministic input. Simple recombinant DNA methods similar to Adleman's experiment can be used. In this manner, we can construct an initial tube containing all initial configurations, with various different nondeterministic choices. By advancing each configuration in parallel and extracting appropriate final configurations after a polynomial number of steps, any NP computation can be implemented using DNA.

3   PSPACE

Adleman proposed to have solved Hamiltonian Path using DNA-based methods [1]. While we have objected that such methods will ultimately fail because they require an exponential amount of material [3, 5], it is interesting to consider the complexity-theoretic implications of such parallelism.

In particular, if we make the presumptuous assumption of massive parallelism needed to support Adleman's and Lipton's NP computations, then in fact, NP is a vast understatement of the power of biocomputers. We show that PSPACE is "feasible."

A $S(n)$ space-bounded TM can enter at most $N = |Q|S(n)|\Sigma|^{S(n)}$ configurations without cycling. To detect whether such a machine accepts its input (and what its result is, if so), it suffices to construct a tube containing $\{C\#\delta^N C\}$. By extracting those molecules containing $C_0(x)$ in their first half, where $C_0(x)$ is the initial configuration on input x, we can determine $\delta^N C_0(x)$. Note that $\log N$ is polynomial in N, thus our parallelism is still within the range of the presumptuous assumption used for NP.

To construct this "lookup-table," we employ a form of parallel pointer-jumping, implemented as a special-case form of communication. The results of 2 permit us to construct a tube containing $\{C\#\delta C\}$ for all configurations C. Using $\lceil \log_2 N \rceil$ phases, we derive a tube containing $\{C\#\delta^{\hat{N}} C\}$, where $\hat{N} = 2^{\lceil \log_2 N \rceil}$.

Each phase produces $T_{2i} = \{C\#\delta^{2i}C\}$ from $T_i = \{C\#\delta^i C\}$. There are a few particular ways to implement this, but we illustrate just one. For simplicity, we address the similar goal of producing $T_{2i} = \{C\#\delta^{2i}C \Diamond C\#\delta^{2i}C\}$ from $T_i = \{C\#\delta^i C \Diamond C\#\delta^i C\}$, where $\Diamond$ is a delimiter containing an EcoRI restriction enzyme site. Let $\lhd$ and $\rhd$ denote the results of cutting $\diamond$.

First, (super-)quadratically increase the populations in T_i using PCR. Cut the results with EcoRI, producing $\{C\#\delta^i C \lhd\} \cup \{\rhd D\#\delta^i D\}$. Allow to recombine, producing a new tube $U = \{C\#\delta^i C \diamond D\#\delta^i D\}$ for all configurations C, D. (Note that the increase in populations ensures that all combinations are represented.) Apply a Turing machine to U that compares $\delta^i C$ to D and overwrites $\diamond$ with ! if $\delta^i C \neq D$. Otherwise, the machine does some copying to give $C\#\delta^i D \diamond C\#\delta^i D$; note that $\delta^i D = \delta^{2i} C$. Discard those molecules containing !; the remaining molecules are of the desired form.

(This represents a primitive form of communication, in that the messages [right-hand halves] are dispersed and arrive at various destinations. Only those combinations in which the message has arrived at a correct destination are retained. The "source" molecules [left-hand halves] are amplified so that those which receive an incorrect message can be discarded. Here, the "addresses" are simply the configurations, D. Better refinements are possible.)

4   Discussion

One appealing aspect of our design for a universal computer is that it proposes a novel implementation of context-specific, site-directed mutagenesis. Despite a 10-year hiatus from the field of molecular biology, the author would comfortably bet that these editing methods are reasonably feasible (for reasonably-sized oligonucleotides, say between 10 and 500 in length). This may represent a contribution independent of molecular computing; and future studies of molecular computing may also provide novel techniques for molecular biology, even if biocomputers remain little more than a purely intellectual exercise.

On the optimistic side, our results do suggest that a fairly direct form of universal computation may be feasible. We have characterized the power of biocomputing in terms of classical complexity theory, obtaining not just NP but PSPACE (or in actual practice, BPP).

In contrast to methods based on Adleman's techniques, including Lipton's further refinements, we do not require a "generation of diversity" step to perform highly parallel computation. Generation of diversity may be characteristic of NP (and PSPACE) solutions, but this does not mean it is desirable, acceptable, or necessary in other cases.

We note that Smith and others have suggested that the PSPACE result uses greater time than the NP methods, in the sense that it requires exponential time for the matching to occur [14]. In part, such commentary arises from a misunderstanding of our proposal. In the pointer-jumping step, it is not necessary that each molecule seek out an appropriate partner from among a large pool, thereby requiring a large amount of time for this step. Rather, we require that the tube be mixed thoroughly before recombination occurs in the pointer-jumping step. After such mixing, the recombinations are local and "immediate." Although the vast majority of recombinations are fruitless, the polynomial increase in population gives rise to $\Omega(1)$ successful recombinations for each original molecule.

In comparison, a separation step *à la* Adleman or Lipton requires filtration or mixing of an exponential (2^{poly}) volume. Thus, the presumptuous assumptions (2^{poly} parallelism, 2^{poly} mixing in one step) needed to achieve NP do suffice to achieve PSPACE, without requiring additional presumption. (We note that further refinements enable the communication and pointer-jumping to be executed without increasing the material used, but mixing remains a problem.)

On the skeptical side, such presumption is difficult to defend. Even ignoring the scaling problems, executing a large number of steps with minimal loss and error is a more tremendous task than building a Pentium from vacuum tubes.

On the 7-city scale solved by Adleman, however, a proof-of-concept experiment may be feasible. And the unexpected spin-offs for molecular biology – unrelated to computation – make this an appealing and potentially fruitful area for collaboration between computer scientists and molecular biologists.

Bibliography

[1] L. Adleman, "Molecular Computation of Solutions to Combinatorial Problems," *Science* **266**, November 1994, 1021–1024.

[2] L. Adleman, "On Constructing a Molecular Computer," manuscript, 11 January 1995.

[3] D. Beaver, "Factoring: The DNA Solution," *Advances in Cryptology – Asiacrypt '94 Proceedings*, to appear, Springer-Verlag, 1995, 1994.

[4] D. Beaver, "Molecular Computing," manuscript, 5 December 1994.

[5] D. Beaver, "Computing With DNA," submitted 12 December 1994 to *Journal of Computational Biology*, and to appear, **2**:1, 1995, pp. 1–7.

[6] D. Beaver, "Molecular Computing," CSE-95-001, 31 January 1995.

[7] D. Boneh, C. Dunworth, R. Lipton, "Breaking DES Using a Molecular Computer," 1995.

[8] D. Gifford, "On the Path to Computing with DNA," *Science* **266**, November 1994, 993–994.

[9] J. Hopcroft, J. Ullman, "Introduction to Automata Theory, Languages, and Computation." Addison-Wesley, 1979.

[10] R. Lipton, "Speeding Up Computations via Molecular Biology." Princeton Technical Report, 1994.

[11] R. Lipton, "Using DNA to solve NP-complete Problems," Princeton Technical Report, 1995.

[12] J. Reif, "Parallel Molecular Computation," Manuscript, 2/22/95, extended abstract to appear, SPAA '95.

[13] R. Rivest, A. Shamir, L. Adleman, "A Method for Obtaining Digital Signatures and Public Key Cryptosystems," *Communications of the ACM* **21**:2, 1978, 120–126.

[14] W. Smith, "An opinionated, but reasonably short, summary of the Mini DIMACS Workshop on DNA based computers, (held at Princeton University on April 4 1995)," 5 April 1995.

[15] W. Smith, A. Schweitzer, "DNA computers in vitro and vivo." NEC Tech Report, 20 March 1995.

Also see the online bibliography at

`http://www.transarc.com/` ~`beaver/research/alternative/molecute/molec.html`

DIMACS Series in Discrete Mathematics
and Theoretical Computer Science
Volume **27**, 1996

Breaking DES Using a Molecular Computer

Dan Boneh   Christopher Dunworth
dabo@cs.princeton.edu   ctd@cs.princeton.edu

and

Richard J. Lipton[1]
rjl@cs.princeton.edu

Department of Computer Science
Princeton University
Princeton, NJ 08544

Abstract

Recently Adleman [1] has shown that a small traveling salesman problem can be solved by molecular operations. In this paper we show how the same principles can be applied to breaking the Data Encryption Standard (DES). Our method is based on an encoding technique presented in Lipton [5]. We describe in detail a library of operations which are useful when working with a molecular computer. We estimate that given one arbitrary (plain-text, cipher-text) pair, one can recover the DES key in about 4 months of work. Furthermore, if one is given cipher-text, but the plain text is only known to be one of several candidates then it is still possible to recover the key in about 4 months of work. Finally, under chosen cipher-text attack it is possible to recover the DES key in one day using some preprocessing.

1 Introduction

Due to advances in molecular biology it is nowadays possible to create a soup of roughly 10^{18} DNA strands that fits in a small glass of water. Adleman [1] has shown that each DNA strand can be used to perform computations. Thus, a small test tube containing DNA

---

[1]Supported in part by NSF CCR–9304718.

strands seems to have more computing power than the most powerful parallel computers. The drawback of this approach is that basic operations using DNA take a long time (e.g. 3 hours). Thus, we are capable of performing 10^{18} basic operations at once, though each operation requires several hours to complete. Throughout the paper we will refer to operations done on DNA strand as bio-steps. This will be explained in more detail in Section 2.

Recently, Lipton [5] has come up with an encoding scheme that enabled him to solve the satisfiability problem of formulas with a small number of variables. A generalization of this scheme [3] can be used to solve the satisfiability problem for circuits. Using these methods, the number of bio-steps required to find a satisfying assignment is proportional to the size of the circuit. Lipton's approach raises the hope of using molecular computing to solve hard problems that come up in practice. However, to make this procedure practical one needs to encode the problem at hand so as to minimize the number of bio-steps required to solve it.

In this paper we give an example of a hard problem that can be solved using a molecular computer. We present a "molecular program" for breaking the Data Encryption standard [7] or DES for short. We precisely describe the steps needed to implement this scheme and show efficient ways of reducing the number of total bio-steps required.

DES is a widely used encryption procedure. It encrypts 64 bit messages and uses a 56 bit key. By breaking DES we mean that given one (plain-text, cipher-text) pair we can find a key mapping the plain-text to the cipher-text. In fact, we can do even better. Suppose we were able to intercept some cipher-text, but we do not know what the plain-text is. What we do know is that the plain-text is one of several possible candidates. Then we can recover the set of keys mapping the candidate plain-texts to the intercepted cipher-text for the same amount of work as when the plain-text was known. Finally, using chosen cipher text attack[2] it is possible to recover the key in one day of work using some preprocessing. This means that after the preprocessing work is done it is possible to break many DES systems for very little work.

Recently, several researchers [16, 2, 8] have come up with various methods for implementing non-deterministic Turing Machines using molecular computers. Clearly a non-deterministic Turing Machine

---

[2]Chosen cipher text attack means that we can obtain a (plain-text, cipher-text) pair where we get to choose the cipher-text.

can break any crypto-system, including DES, by guessing the correct key. Though these results are very important theoretically, they are not useful in practice. For instance, breaking DES using a Turing Machine would require millions of biological operations. Running such experiments will take hundreds of years.

It should be pointed out that attacks using differential cryptanalysis methods have proven to be very useful for breaking DES [4, 6]. The central achievement of these attacks is that DES can be broken on a conventional computer in 2^{43} steps. This means that DES can be broken within several days on a conventional computer. Note however, that the differential cryptanalysis methods require 2^{43} pairs of (plain-text, cipher-text), while our molecular computer requires only one such pair. In fact, as was mentioned above, a molecular computer can break DES with even less information.

Another conventional attack on DES was suggested by Wiener [10]. Wiener predicted that using dedicated hardware which costs $1 million it is possible to run through all 2^{56} DES keys in 7 hours. An advantage of the molecular attack is that the plain-text need not be known exactly. Furthermore, if DES was made to use 64 bit keys, the conventional attack would slow down by a factor of 256, while the molecular attack would be uneffected. Furthermore a molecular computer is likely to be substantially cheaper.

Throughout this paper we will assume an error free model. We assume that all the molecular biology experiments work perfectly without any errors. Though this assumption does not hold in practice, it will do for now since the objective of this paper is to demonstrate that molecular computers have the potential of solving interesting real world problems.

In Section 2 we introduce the basic functions of a molecular computer. We then go on to briefly sketch the DES circuit in Section 3. Sections 4 and 5 describe the main steps of the procedure.

2   Overview of molecular computing

In order to understand the computation described here, it is necessary to gain a fundamental understanding of DNA's structure and function. The reader who is familiar with this information may wish to skip to Section 2.1 which covers our notation.

DNA (deoxyribonucleic acid) is found in every living creature as the storage medium for genetic information. It is comprised of subunits called nucleotides that are strung together into polymer chains (the chemical structure of a single nucleotide is shown in Figure 1. DNA polymer chains are more commonly called DNA *strands*, and short strands are called *oligonucleotides*, or simply *oligos*.

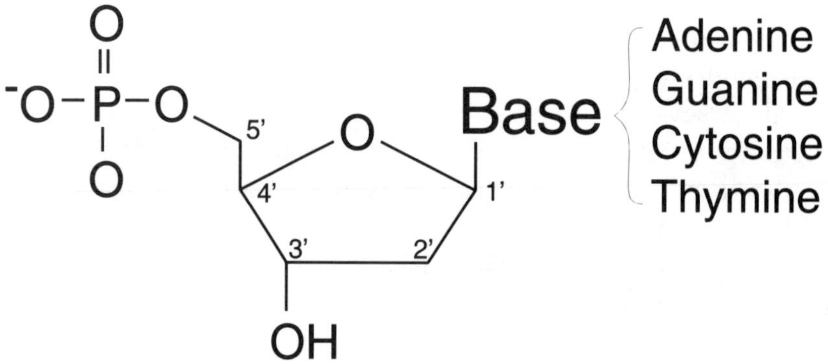

Figure 1: Chemical structure of a nucleotide

There are four kinds of nucleotides in DNA, distinguished by the chemical group, or base, attached to it. The four bases are adenine, guanine, cytosine, and thymine, abbreviated as A, G, C, and T (we will henceforth use these letters to refer to nucleotides containing these bases). Single nucleotides are linked together end-to-end to form DNA strands in a process called polymerization. This linking occurs via a reaction between the 5' phosphate of one nucleotide and the 3' hydroxyl of another (see Figure 2). Note that every DNA strand will have two distinct ends—one with a free 5' PO_4 group and the other with a free 3' OH group, referred to as the 5' and 3' ends, respectively. Thus, we can think of the sequence of nucleotides in any strand as having a natural orientation. This orientation is important to remember because it restricts the kinds of operations that one can perform on the strands.

DNA does not usually exist in nature as free single strands, though. Under appropriate conditions single strands will pair up and twist around each other, forming the famous double helix structure discovered by Watson and Crick. This pairing occurs because of a mutual attraction, called hydrogen bonding, that exists between As and Ts and between Gs and Cs. Hence, A/T and G/C are called *complementary* base pairs. A further condition is that hydrogen

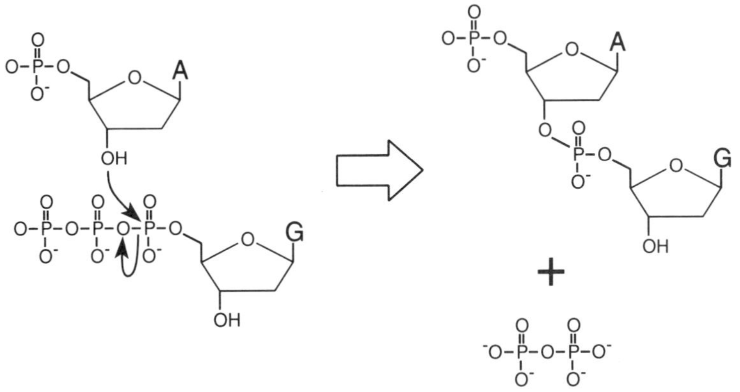

Figure 2: DNA polymerization reaction

bonding will only occur if the pair of complementary strands are oriented in an *antiparallel* fashion. That is, as one strand of a double helix runs in the 5' → 3' direction, its complement runs in the 3' → 5' direction. For example, if we have the strands 5'-AAGCGTAG-3', 3'-TTCGCATC-5', and 5'-TTCGCATC-3', the first two can pair up into a double strand—namely, ⁵'⁻ᴬᴬᴳᶜᴳᵀᴬᴳ⁻³' —but the first and last ones cannot, because when lined up in an antiparallel fashion we see that they are not complementary: ⁵'⁻ᴬᴬᴳᶜᴳᵀᴬᴳ⁻³' . This process of complementary strands coming together to form a double helix is called *annealing*, and the reverse process—a double helix coming apart to yield its two constituent single strands— is called *melting*.

Of course, DNA strands would not be of much use to us unless we had some way to manipulate them. To accomplish this we use any of a number of commercially available enzymes that perform the tasks we require. One class of enzymes, called restriction endonucleases, will recognize a specific short sequence in a strand and then "cut" the strand at that location. Another class of enzymes, the DNA polymerases, will read a single DNA strand, called the "template" strand, and create its complementary strand. Still another enzyme, called DNA ligase, will hook together, or ligate, the free 5' end of one strand to the free 3' end of another strand. There are many other enzymes that could potentially be useful, but for our computation these are sufficient.

2.1   Notation

We have devised a notation for abstractly representing DNA strands in order to succinctly describe the changes they undergo during various operations.

First, we represent strings over the alphabet {A,C,G,T} using letters for variable names (e.g. x = ACCTGAC). Note that x is a *string*, but not yet a *strand*—that is, it's just a sequence of characters, with no orientation implied. We indicate the concatenation of two or more strings by juxtaposition (e.g. if y = AAATAAG, then xy = ACCTGACAAATAAG).

Two operations that we would like to apply to strings are complement and reverse. The *complement* of a string x, denoted $\hat{x}$, is the string that results when each character of x has been replaced by its Watson-Crick complement. That is, we apply the following mapping to every character in x: A→T, C→G, G→C, T→A. Thus, for x = ACCTGAC, we have $\hat{x}$ = TGGACTG. The *reverse* of a string x, denoted x^R, is just the string x in reverse order. Thus for the same example, x^R = CAGTCCA.

DNA strands are more than just strings, though—they are strings *with an orientation*. To this end we introduce some new operators that, when applied to a string, denote a DNA strand based on that string.

The first such operator is "↑". ↑ x denotes the single DNA strand whose sequence is x and whose orientation is $5' \rightarrow 3'$, as x is read from left to right. For example, if x = ACCTGAC, then ↑ x is the strand 5'-ACCTGAC-3'. The second such operator is "↓". ↓ x denotes the strand complementary to ↑ x. For our example, ↓ x = 3'-TGGACTG-5'. Finally, we have the operator "↕". ↕ x denotes the double strand that results when ↑ x and ↓ x anneal in solution.

The table below summarizes these operators:

$$x \quad = \quad \text{ACCTGAC}$$
$$y \quad = \quad \text{AAATAAG}$$

$$xy \quad = \quad \text{ACCTGACAAATAAG}$$
$$\hat{x} \quad = \quad \text{TGGACTG}$$
$$x^R \quad = \quad \text{CAGTCCA}$$
$$\hat{x^R} \quad = \quad \text{GTCAGGT} \qquad (= \hat{x}^R)$$

$$\uparrow x \quad = \quad \text{5'-ACCTGAC-3'}$$
$$\downarrow x \quad = \quad \text{3'-TGGACTG-5'} \qquad \text{(note: } \uparrow \hat{x}^R = \text{5'-GTCAGGT-3'} = \downarrow x)$$
$$\updownarrow x \quad = \quad \begin{matrix} \text{5'-ACCTGAC-3'} \\ \text{3'-TGGACTG-5'} \end{matrix}$$

The convenience of this notation becomes clearer when we try to describe various operations on DNA strands in solution. For example, we can easily see that the strand $\uparrow xy$ and the strand $\downarrow yz$ will anneal under appropriate conditions to form the expected duplex $\uparrow x \updownarrow y \downarrow z$. We will rely on this notation to describe aspects of our computation as appropriate.

2.2   Biological operations

Our fundamental model of computation will be to apply a sequence of operations to a set of strands in a test tube. The operations that we make use of are derived from a collection of experiments commonly used in molecular biology today [1].

2.2.1 Extract

We will need the ability to extract from a test tube all strands that contain any specific short nucleotide sequence. To accomplish this we will use the method of biotin-avidin affinity purification as described in [1]. This technique works in the following way. If we want to extract all strands containing the sequence $\uparrow x$, then we first create many copies of its complementary oligo, namely $\downarrow x$. To these oligos we attach a biotin molecule, which will in turn be anchored to an avidin bead matrix. If we then melt the strands in our test tube and pour them over this matrix, those strands that contain $\uparrow x$ will anneal to the $\downarrow x$ oligos anchored to the matrix. A simple wash procedure will whisk away all strands that did not anneal, leaving behind only those strands that contain $\uparrow x$, which can then be retrieved from the matrix. We refer to this operation as an *extract* using beads of type $\downarrow x$.

2.2.2 Polymerization via DNA Polymerase

Given a particular single strand of DNA, we may wish to create its Watson-Crick complementary strand. To do this we will use the enzyme DNA polymerase. DNA polymerase will "read" the given strand, called the template strand, in the $3' \rightarrow 5'$ direction and build the complementary strand in the $5' \rightarrow 3'$ direction, one nucleotide at a time. In order to work, DNA polymerase actually requires that there be a short portion of the template that is double stranded, and it is onto the end of this short complementary piece, called the *primer*, that the enzyme will add the new nucleotides. For example, if we have some strand $\uparrow xyz$, DNA polymerase cannot create its complement. However, if we add $\downarrow z$ to the solution and let it anneal to $\uparrow xyz$, we will have $\uparrow xy \updownarrow z$, and DNA polymerase will be able to add nucleotides onto the free 3' end of z to create $\updownarrow xyz$. Note that because DNA polymerase only works in one direction, the partial duplex $\uparrow x \updownarrow y \uparrow z$ will yield $\updownarrow xy \uparrow z$ and not the full duplex $\updownarrow xyz$.

2.2.3 Amplification via PCR

At times we will need to make copies of all the DNA strands in a test tube. This can be done with a straightforward application of the polymerase chain reaction (PCR). PCR is a process that uses DNA polymerase to make many copies of a DNA sequence that exists between two primer sequences. PCR works in the following way. If we have the duplex $\updownarrow xyz$, we first melt it to form $\uparrow xyz$ and $\downarrow xyz$. To this solution we will add the primer oligos $\downarrow z$ and $\uparrow x$, which anneal to form the partial duplexes $\uparrow xy \updownarrow z$ and $\updownarrow x \downarrow yz$. DNA polymerase can then elongate the primers to create full duplexes of the form $\updownarrow xyz$. Note that we now have *two* copies of our original strand. If we just repeat this process, we will again double the number of copies of the original strand in solution. Soon we will have four copies, then eight, then sixteen, and so on, until we have enough copies for our purposes. Thus, if we can guarantee that the primer sequences that we use occur on the ends of every strand, and only on the ends, then we can use PCR to duplicate every strand in the test tube. We will call this operation *amplify*.

3   The DES circuit

In this section we give a brief overview of the DES circuit. DES encrypts a 64 bit plain-text into a 64 bit cipher-text using a 56 bit key. We denote by $\mathrm{DES}(M, k)$ the DES encryption of the plain-text M using the key k.

As will be seen in the next sections, our goal is to run the DES circuit on a fixed 64 bit string M using all possible keys k. In other words, we plan to evaluate the function $f(k) = \mathrm{DES}(M, k)$ for all possible k. The circuit computing the function $f(k)$ is shown in figure 1. The figure displays a convoluted version of the DES circuit since in our case the message to be encrypted is fixed and the key varies. Traditionally the circuit is drawn the other way around, i.e. the key is fixed and the message varies. For a more detailed description of the DES circuit see [7].

We now explain the various components of figure 3. The circuit is composed of 16 levels. We refer to each level as a round. The figure shows the first four rounds and the last round. The input to the circuit is the 56 bit key shown on the left. The high order 32 bits of M are denoted by M_h and the low order bits are denoted by M_l.

A P-box is a box which permutes the bits of its input. Suppose the input to a P-box contains x bits and the output contains y bits. If $x = y$ then the box will only permute the bits of the input, e.g. the box may swap the second and third bits, mapping 010 to 001. If $x > y$ then the box will output a subset of the bits of the input in some order. If $x < y$ then the box will replicate some of the bits of the input to obtain a y bit number. The exact operation performed by each P-box in the figure is predetermined and can be found in [7]. For our purposes we note that the P-boxes are insignificant and may be ignored. The reason for this is that P-boxes simply change the order of the bits that arise during the computation. When going through a P-box there is no need to physically change the order of the bits in the DNA-strand. We can simply keep track in our mind which bits on the strand encode which bits of the computation.

An S-box is more complicated. An S-box takes 48 bits of input and outputs 32 bits. The S-box groups the 48 input bits into 8 groups of 6 bits each. For each group of 6 bits the S-box performs a table lookup and outputs 4 bits. This can be seen in figure 4. The 8 tables used can be found in [7]

To summarize, we see that each DES round involves a 48 bit Xor, an S-box and another 32 bit Xor. Thus, each round requires 80 Xor operations and 8 table lookups.

A useful property of DES is that the decryption procedure is very similar to the encryption procedure. We denote the decryption procedure by DES^{-1}. That is, if $E = \text{DES}(M, k)$ then $M = \text{DES}^{-1}(E, k)$. The only difference between the DES circuit and the DES^{-1} circuit is in the P-boxes. Thus, as far as our biology experiment is concerned the two circuits may be regarded as the same circuit.

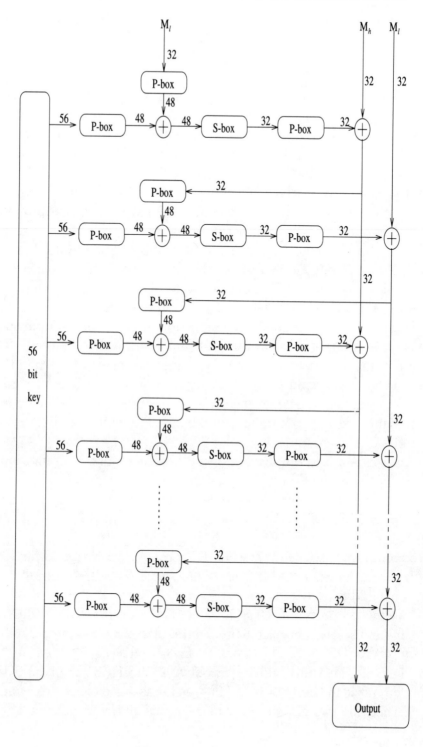

Figure 3: DES circuit

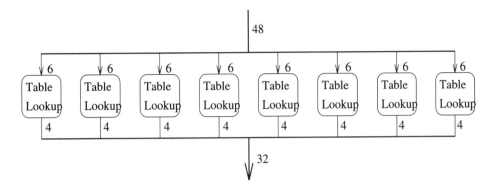

Figure 4: S-box circuit

4 Representing binary strings

Before we explain how to evaluate the DES circuit we first describe a convenient method for encoding binary strings as DNA strands. The encoding is based on the one described in [5].

Let $x = x_1 \ldots x_n$ be an n-bit binary string. The idea is to assign a unique 30-mer to each bit position and bit value.

1. Let $B_i(0)$ be the 30-mer used to encode the fact that the i-th bit of x is 0. Similarly, let $B_i(1)$ be the 30-mer used to encode the fact that the i-th bit is 1.

2. For $i = 0, \ldots, n$ let S_i be a 30-mer which will be used as a separator between consecutive bits.

The DNA strand representing the binary string $x \in \{0, 1\}^n$ will be:

$$\updownarrow S_0 \; B_1(x_1) \; S_1 \; B_2(x_2) \; S_2 \, , \ldots, \; S_{n-1} \; B_n(x_n) \; S_n$$

There are several points to be made regarding this encoding. First of all, it is crucial that the strings $B_i(x)$ and S_i should all be distinct. Furthermore, it is desirable that any pair of these strings should not contain long common substrings. For example, there shouldn't be a string of length 15 which is a substring of both S_0 and S_1. This condition will be satisfied with sufficiently high probability if the words are chosen uniformly and independently at random.

Suppose a solution T contains DNA strands representing a set of n-bit binary strings. By definition, all these strands will start with S_0 and end with S_n. This is convenient for doing PCR, since S_0, S_n can be used as primers.

A potential problem may arise if one strand contains two portions which are Watson-Crick complements of one another. Such a single stranded DNA molecule is likely to anneal to itself, making it useless for our purposes. A simple way of avoiding this problem is to choose the strings $B_i(x)$ and S_i to only contain the bases A and C. Since A and C are not Watson-Crick complements, a strand encoding an n bit number will never contain two complementary portions.

Finally, it is worth noting that the strings S_i are not essential to the encoding. It is possible to represent an n-bit number as $\updownarrow B_1(x_1)\, B_2(x_2)\, \ldots B_n(x_n)$, however it is more convenient to encode the strings using the separators.

Notation: For notational convenience, given an n-bit string x, we denote by $R_i(x)$ the string encoding x at position i, i.e.

$$R_i(x) = B_i(x_1)\, S_i\, B_{i+1}(x_2)\, S_{i+1}\, \ldots S_{i+n-2}\, B_{i+n-1}(x_n)\, S_{i+n-1}$$

4.1   Operations on binary strings

We now explain the basic primitive operations that can be performed on binary strings using their encoding as DNA strands. Throughout this section we let T be a test tube containing a collection of DNA strands which represent some binary strings.

As was discussed in section 2, it is possible to extract all the strands in T that contain a certain sequence of basis. This will be the fundamental operation used for evaluating circuits.

Suppose we wish to extract all strands in T representing binary strings whose i'th bit is 1. We know that all such strands contain the string $B_i(1)\, S_i$ and they are the only strands containing this string. Hence, by extracting all strands in T containing the sequence $B_i(1)\, S_i$ we extract exactly those strings whose i'th bit is 1. We denote this operation by

$$\text{Extract}(\, T\ ,\ \ x_i = 1)$$

It is possible to perform a more general extract operation. Suppose we wish to extract all strands in T who have a 1 in their i'th position and a 0 in their $i+1$'st position. We denote such an operation by

$$\text{Extract}(\, T\ ,\ \ x_i x_{i+1} = 10)$$

Clearly this can be done by extracting all strands containing the string

$$R_i(10) = B_i(1)\, S_i\, B_{i+1}(0)\, S_{i+1}$$

Hence, we were able to perform an "and" operation on consecutive bits for the cost of one extract operation. Note that the fact that the bits are adjacent is crucial for this to work.

An even more general extract operation is the following:

$$\text{Extract}(\ T\ ,\ \ x_i x_{i+1} x_{i+2} = 101 \ \ \text{or} \ \ x_i x_{i+1} x_{i+2} = 001)$$

This can be done using two types of beads at once. One bead will attract all strands containing the string $R_i(101)$ and the other will attract all strands containing $R_i(001)$. Hence, for the cost of one extract operation we were able to perform a combination of an "and" and an "or". For technical reasons it is only possible to perform this simultaneous "or" operation when the strings being extracted have the same length. For instance, in the above example, both strings 101 and 001 have the same length 3.

Important Note: An important observation is that these operations on DNA strands work no matter where the the bits are on the strand. For instance, the string 01 can be represented in two equivalent ways:

$$\updownarrow S_0\ B_1(0)\ S_1\ B_2(1)\ S_2 \qquad \text{or} \qquad \updownarrow S_0\ B_2(1)\ S_2\ B_1(0)\ S_1$$

All we did was to put the string representing the second bit in the first position. Clearly, the Extract$(T, b_2 = 1)$ operation works exactly as before. The point is that the fact $b_2 = 1$ is encoded by the presence of the string $B_2(1)\ S_2$ on the strand. It makes no difference where this string is on the strand. Finally we note that the string 01 can be represented in a third way as:

$$\updownarrow S_2\ B_2(1)\ S_0\ B_1(0)\ S_1$$

This representation will be useful in Section 6.2

5 Plan of DES attack

Now that we know how to represent binary strings and how to perform primitive operations on them, we are ready to explain our attack on DES.

In the next section we will show that given a message M it is possible to create a solution that contains for each $k \in \{0,1\}^{56}$ a DNA strand of the form:

$$\updownarrow S_0\ R_1(\ k\ )\ R_{57}(\ \text{DES}(M,k)\ )$$

Each strand in this solution encodes a key k and the encryption of the fixed message M using the key k. We denote this solution by $\mathrm{DES}(M)$. As was stated in Section 3, the decryption circuit DES^{-1} may be regarded as the same circuit as DES. Thus, we can just as easily create a solution $\mathrm{DES}^{-1}(E)$ which is defined analogously to $\mathrm{DES}(M)$, i.e. contains all strands of the form

$$\updownarrow \ S_0 \ R_1(\ k\ ) \ R_{57}(\ \mathrm{DES}^{-1}(E, k)\ )$$

Let (M, E) be a (plain-text, cipher-text) pair. We wish to find a key k s.t. $M = \mathrm{DES}^{-1}(E, k)$, i.e. the decryption of the cipher-text E is the plain-text M. This can be done as follows:

1. Create the solution $\mathrm{DES}^{-1}(E)$ described above.

2. Extract from $\mathrm{DES}^{-1}(E)$ all strands that contain the pattern $R_{57}(\ M\ )$.

3. The extracted strands encode pairs of strings (k, M) where $M = \mathrm{DES}^{-1}(E, k)$. The key k can easily be recovered by sequencing any of the extracted DNA strands.

Step 2 and 3 of this process can be done very quickly, i.e. within a few days. The laborious part is step 1.

The same procedure can be used even if the plain-text is only known to be one of a number of candidates. Simply repeat step 2 for each one of the candidates. The running time of the process is hardly effected by this since most of the work is done in step 1. Thus, suppose we are able to intercept one cipher-text and we have an idea of what the encoded message is. Then the above procedure will enable us to recover the key and eavesdrop on the rest of the conversation.

Once we generate the solution $\mathrm{DES}^{-1}(E_0)$ for some fixed 64 bit word E_0, we can very quickly break any DES system if we can use chosen-cipher-text attack. Simply generate the pair (M_0, E_0) and run through steps 2 and 3 above. Clearly the same is true if we can use chosen-plain-text attack. Thus, potentially, a chemical company can spend the time to generate the solution $\mathrm{DES}^{-1}(E_0)$ for some fixed E_0 and then sell it to anyone who wants to quickly break a DES system.

6   Evaluating the DES circuit

In this section we describe how to generate the solution $\mathrm{DES}(M)$. As we saw in the previous section, this solution is the key ingredient in the molecular attack on DES.

We give a quick outline of how to construct this solution and then describe the details. The process is carried out in three steps.

1. Construct an initial solution encoding all 56-bit strings. Thus, the solution contains all strands of the form

$$\updownarrow S_0 \ R_1(k)$$

for all $k \in \{0,1\}^{56}$.

2. Evaluate the gates of the DES circuit by tagging on their values. For instance, suppose the first gate in the DES circuit tells us to Xor bit 3 and bit 7 of the key. We do this by appending the value of $k_3 \oplus k_7$ to each of the strands in the initial solution. The resulting solution will contain strands of the form:

$$\updownarrow S_0 \ R_1(k) \ B_{57}(k_3 \oplus k_7) \ S_{57}$$

for all $k \in \{0,1\}^{56}$ where k_3 and k_7 are the third and seventh bits of k.

3. We evaluate the gates in the circuit one by one. When we are done, all strands in the solution look like:

$$\updownarrow S_0 \ R_1(k) \ R_{57}(\ I \ ) \ R_r(\ \mathrm{DES}(M,k) \ )$$

where I is a string of bits corresponding to values generated by internal gates of the circuit.

6.1   Constructing the initial soup

We begin the detailed discussion by explaining how to construct the initial solution of 2^{56} DNA strands corresponding to each of the possible DES keys. We use a procedure similar to the one described in [5].

The basic idea is to use the graph G shown in figure5. Each path from S_0 to S_{56} corresponds to one DES key. To create a soup of DNA strands which correspond to all paths in G we use the encoding of [1]. We follow the following steps:

1. Recall that each vertex in the graph represents a string of length 30. Denote the string assigned to vertex V by $V_l V_r$ where V_l is the 15 left letters and V_r are the 15 right letters.

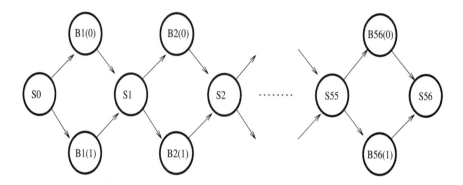

Figure 5: Initialization graph

2. For each vertex v create a test tube containing the strands $\uparrow V_l V_r$. For each directed edge (u, v) create a test tube containing the strands $\downarrow U_r V_l$. To handle the end points, create two test tubes containing $\uparrow (S_0)_l$ and $\downarrow (S_{56})_r$.

3. Mix all the test tubes and allow the strands to anneal and ligate. Then extract from the resulting mixture all the strands that contain the words S_0. Call the set of extracted strands A. From the solution A extract all strands that contain the word S_{56}.

It is not difficult to see that at the end of step 2 we get a mixture that corresponds to various paths in the graph. At the end of step 3 the resulting strands will correspond to all paths from S_0 to S_{56}.

6.2 Tagging

Before we explain how the various components of the DES circuit are evaluated we must first explain how to append a given string to all strands in a test tube. This "append to end" primitive can be implemented in various ways. Here we describe one possible method.

Suppose T' is a test tube containing DNA strands which encode various n bit binary strings. Since all the strings have the same length n, all the DNA strands will end with word S_n.

We explain the tagging procedure through an example. Suppose we wish to append the string 01 to all strings in T'. Formally, this means that we append the DNA strand

$$R_{n+1}(\, 01 \,) = B_{n+1}(0) \, S_{n+1} \, B_{n+2}(1) \, S_{n+2}$$

to all strands in T'. We refer to this operation as tagging the strands with 01. To do this, we create a test tube T_{01} containing many copies of the single stranded DNA:

$$\downarrow S_n \; R_{n+1}(\,01\,)$$

We then pour the contents of T_{01} into T'. Since all the strands in T' end with S_n the new strands will anneal to them. The resulting strands will look like:

$$b_1 \; b_2 \; \ldots \; b_{n-1} \; b_n \;\; S_n$$
$$S_n \; B_{n+1}(0) \; S_{n+1} \; B_{n+2}(1) \; S_{n+2}$$

We can now apply a polymerase enzyme to complete this DNA molecule to a double stranded molecule, as explained in Section 2.2.

6.3 Evaluating a lookup table

It remains to show how to evaluate the various components of the DES circuits. There are only two types of gates in the DES circuit: an Xor gate and a lookup table (used in the S-boxes). We begin by explaining how to evaluate a lookup table.

A lookup table is a function $f : \{0,1\}^6 \to \{0,1\}^4$ which maps 6 bit strings to 4 bit strings. The function is given as a table of values. For example,

	0	1	2	3	4	5	6	7	8	9	10	11	12	13	14	15
0	14	4	12	1	2	15	11	8	3	10	6	12	5	9	0	7
1	0	15	7	4	14	2	13	1	10	6	12	11	9	5	3	8
2	4	1	14	8	13	6	2	11	15	12	9	7	3	10	5	0
3	15	12	8	2	4	9	1	7	5	11	3	14	10	0	6	13

Columns correspond to the right 4 bits and rows to the left 2 bits. For example, 2 would be mapped to 12. This look up table is one of the eight tables used in the DES S-boxes.

In the DES circuit it is possible to arrange things so that the 6 bits to be looked up are always consecutive bits. Say we wish to perform a table look up on the value of the 6 bits in position i to $i+5$. Thus, the strands look like

$$\updownarrow W_l \; B_i(x_i) \; S_i \; B_{i+1}(x_{i+1}) \; \ldots \; B_{i+5}(x_{i+5}) \; S_{i+5} \; W_r$$

where W_l and W_r are the bits on the left and right of the critical 6-bit positions $i, \ldots, i+5$. For a given strand, let $x = x_i \ldots x_{i+5}$ be the bit string encoded on the strand at position i. We wish to append to every strand the 4 bit string $f(x)$ where f is the function computed by the look up table. This is done in two steps:

1. Separate the solution into 16 containers $T_0, \ldots T_{15}$ according to the value to be appended. Formally this entails the following extract operation:

$$T_j \leftarrow \text{Extract}(\, T \,, \; f(x) = j\,)$$

2. For each $j = 0, \ldots 15$, append the 4 bit binary string representing j to all strands in container T_j. Formally, we tag all strands in container j by the string $R_n(j)$ where n is the current length of all strands.

This procedure is best explained through an example. Using the above table we see that container T_0 should contain all strands that have one of the 6-bit patterns $15, 16, 31$ and 45 in position $i, \ldots, i+5$. We can extract all these DNA strands using the following extract:

$$T_0 \leftarrow \text{Extract}(\, T \,, \; x = 15 \text{ or } x = 16 \text{ or } x = 31 \text{ or } x = 45)$$

We then tag all strands in T_0 with the string 0000. Note that all 4 bits can be tagged on at once.

Step 1 of this procedure requires 64 different types of beads corresponding to all 6 bit combinations. These beads can be prepared using the same method used to construct the initial soup of all 56 bit strings, i.e. by constructing all paths in a graph similar to Figure 5 which is made up of 6 diamonds. As we will see in Section 6.2 the same set of 64 different beads can be used for all rounds of the DES circuit. Hence, they need only be prepared once at the beginning of the computation. The same method can be used to generate the 16 bit patterns that are appended to the strands in step 2 of the procedure.

To summarize, we have just seen that a look up table operation can be done using 16 extract operations which may be done in parallel followed by 16 parallel tag operations. Hence, if we allow 16 extracts to be done at once, then table look up can be done in time which is approximately equivalent to one extract operation and one tag operation.

6.4 Evaluating Xor gates

Finally, we explain how to evaluate an Xor gate. Suppose we wish to evaluate the Xor of the i'th and j'th bits of the strings in the current solution T. As usual, this is done by appending the value $x_i \oplus x_j$ to the strands representing x. The first step towards performing this operation is to separate the solution T into two solutions T_0, T_1 where T_0 contains those strands in T satisfying $x_i \oplus x_j = 0$ and T_1 contain those satisfying $x_i \oplus x_j = 1$. We then tag each of T_0, T_1 with the appropriate value.

We can write $x_i \oplus x_j$ as

$$x_i \oplus x_j = (x_i + x_j)(\bar{x}_i + \bar{x}_j) \quad ; \quad \overline{x_i \oplus x_j} = (\bar{x}_i + x_j)(x_i + \bar{x}_j)$$

This leads the following procedure for generating T_0, T_1:

1. Set $T^{11} = \text{extract}(T, x_i = 1 \text{ or } x_j = 1)$ and $T^{01} = \text{extract}(T, x_i = 0 \text{ or } x_j = 1)$.

2. Set $T_1 = \text{extract}(T^{11}, x_i = 0 \text{ or } x_j = 0)$ and $T_0 = \text{extract}(T^{01}, x_i = 1 \text{ or } x_j = 0)$.

This procedure requires 4 extract operations. However, if we allow two extract operations to be done in parallel, then the procedure can be carried out in only two steps.

As we saw in Section 4.1, "and" gates are inherently sequential, while an "or" of many variables can be done in one step. This suggests that when writing out a circuit, one should try to minimize the number of "and" gates in the circuit. This is the reason why we wrote the formulas for Xor as a CNF expression. Had we used DNF expressions there would have been 4 and gates (for the Xor and its complement) as opposed to two in the CNF expression. Consequently, evaluating the Xor using CNF expressions would have required 6 extract steps.

Thus, one Xor gate can be evaluated in time equivalent to two extract steps and one tagging step. If we allow more extract steps to be carried out in parallel, then we can evaluate more than one Xor gate at a time. We discuss this improvement next.

6.5 Evaluating three Xors at once

We show that if one can perform 32 extract operations at once, then it is possible to evaluate three Xor gates in two steps. It is possible to arrange things so that the bits to be Xored are x_i, x_{i+1}, x_{i+2} and x_j, x_{j+1}, x_{j+2}, i.e. the value to be appended to the strands is $x_i \oplus x_j, x_{i+1} \oplus x_{j+1}, x_{i+2} \oplus x_{j+2}$. The point is that we are Xoring two groups of three consecutive bits.

As usual, we plan to separate the solution T into 8 solutions according to the value to be tagged. We show how to do this separation in 2 steps. The first step consists of 8 extracts and the second consists of 32 extracts.

For $z_1 z_2 z_3 \in \{0, 1\}^3$ define

$$Z(z_1 z_2 z_3) = z_1 \, z_2 \, (z_1 \oplus z_2 \oplus z_3) \in \{0, 1\}^3$$

i.e. $Z(z_1 z_2 z_3)$ is a 3 bit string whose third bit is $z_1 \oplus z_2 \oplus z_3$. To separate T into $T_{000}, \dots, T_{111}$ according to the value to be tagged perform the following steps:

1. For each $z_1 z_2 z_3 \in \{0, 1\}^3$ do
$$T_{z_1 z_2 z_3} \leftarrow \text{Extract} \quad (T, \; x_i x_{i+1} x_{i+2} = Z(z_1 z_2 z_3) \text{ or,}$$
$$x_j x_{j+1} x_{j+2} = Z(z_1 z_2 0) \,)$$

2. For each $z_1 z_2 \in \{0, 1\}^2$ and $a = a_1 a_2 a_3 \in \{0, 1\}^3$ do
$$T_{a, z_1 z_2} \leftarrow \text{Extract} \quad (T_{z_1 z_2 b}, \; x_i x_{i+1} x_{i+2} = a \oplus Z(z_1 z_2 b) \text{ or}$$
$$x_j x_{j+1} x_{j+2} = a \oplus Z(z_1 z_2 0) \,)$$
where $b = \overline{a_1 \oplus a_2 \oplus a_3}$.

3. For all $a \in \{0, 1\}^3$ do
$$T_a \leftarrow T_{a,00} \bigcup T_{a,01} \bigcup T_{a,10} \bigcup T_{a,11}$$

For each in $a = a_1 a_2 a_3 \in \{0, 1\}^3$ we can now tag all strands in T_a by the bits $a_1 a_2 a_3$. The reason this extract procedure works follows from the fact that for any $a_1 a_2 a_3 \in \{0, 1\}^3$ we have:

$$x_1 x_2 x_3 \oplus y_1 y_2 y_2 = a_1 a_2 a_3 \Longleftrightarrow$$
$$\exists \, z_1 z_2 \in \{0, 1\}^2 \text{ s.t. } \left(x_1 x_2 x_3 = Z(z_1 z_2 b) \quad \text{or} \quad y_1 y_2 y_3 = Z(z_1 z_2 0) \right)$$
$$\text{and}$$
$$\left(x_1 x_2 x_3 = a \oplus Z(z_1 z_2 b) \text{ or } y_1 y_2 y_3 = a \oplus Z(z_1 z_2 0) \right)$$

where $b = \overline{a_1 \oplus a_2 \oplus a_3}$.

It is worth noting that there is a simple extraction procedure for evaluating three Xors in two steps if we are allowed to perform 64 extractions per step. Simply separate the solution T into 8 solutions according to the value of $x_i x_{i+1} x_{i+2}$ and then separate each of the resulting 8 solutions into 8 solutions according to the value of $x_j x_{j+1} x_{j+2}$. Using this procedure we perform 8 extractions on the first step and 64 extractions on the second step. This should be contrasted with the performance of the procedure given above which only requires 32 extractions on the second step.

6.6   Shortening the strands

The process of continually appending values to the strands would make the strands very long by the time we complete the DES circuit. It is possible to shorten the strands by throwing away segments which are no longer needed. For instance, this can be done at the end of every round. The only values that have to be kept on the strands is the 56 bit key and the 64 bits that propagated out of the previous round. For a strand that looks like $\updownarrow K\ J\ P$ where K is the 56 bit key, J is the junk portion and P are the 64 bits from the previous round that have to be propagated to the next round, we would like to remove that junk portion J.

Removing the middle of a strand is a complicated operation. Fortunately there is a simple fix. As was discussed in Section 4.1, it makes no difference where the string P is on the strands. Hence, when tagging the strands with the bits of P we can place P next to the key K. At the end of the round the strands will look like: $\updownarrow P\ K\ J$. We can now use a restriction enzyme to cut the J portion of the strand. At the beginning of the next round all strands look like $\updownarrow P\ K$.

This procedure shows that cutting the junk portion of the DNA strands is a simple operation, if the data is arranged with foresight. It involves the use of a restriction enzyme and the filtering of the junk strands out of the solution.

Since the B and S strings used by the junk portion are no longer present in the solution they may be reused. For instance, if the 57'th bit was cut away from all strands then the strings $B_{57}(1), B_{57}(0)$ and S_{57} can be reused in the next round. Specifically, the same set of 64 beads used in the table lookup step can be used in all rounds.

6.7   Summary thus far

Before going any further we briefly calculate the number of steps required to evaluate the DES circuit using the methods sketched thus far. First, observe that the first $48 + 32 + 32 = 112$ Xors may be ignored. These Xors are computing an Xor with the message bits. Since the message bits are fixed, these Xors are either a no-op (i.e. no operation performed) or a negation. A negation only requires that we make a mental note remembering that the string representing 1 is now representing 0 and vice versa.

Suppose we allow 32 parallel extractions per step. We saw that it is possible to evaluate 3 Xors in two parallel extraction steps and one table lookup per step. The group of 48 Xors on the left of the DES circuit have to be evaluated in 15 of the rounds. Evaluating 48 Xors takes $\lceil 48 * \frac{2}{3} \rceil$ parallel extraction steps. Similarly the group of 32 Xors on the right of the circuit must be evaluated in 14 of the rounds. In addition there are 128 table lookups in the DES circuit (8 per round). Hence, the total number of parallel extractions required is

$$\left[15 * \lceil 48 * \frac{2}{3} \rceil + 14 * \lceil 32 * \frac{2}{3} \rceil \right] + 128 = 916$$

We can only roughly estimate the amount of lab time required to run this algorithm. Assuming 10 extract operations per day, this experiment would require roughly 4 months of lab work. Clearly many factors have been excluded form this estimate and it should only be regarded as a rough approximation.

7   Speeding computations using Join

We now describe a general method which enables us to perform more steps in parallel. The major bottleneck of the algorithm described in Section 6 is the evaluation of Xors. When the algorithm had to evaluate 32 Xors, it had to evaluate them one by one (or in groups of three). A natural thing to try is to make two copies T_1, T_2 of the solution T. Then evaluate the left 16 Xors in T_1 and the right 16 Xors in T_2 and then combine the solutions T_1 and T_2. Unfortunately there is no simple way of combining the solutions T_1 and T_2 after the Xors have been evaluated.

The problem can be stated as follows. Suppose a strand representing the string x is present in the solution T. We wish to compute two functions $f(x)$ and $g(x)$ and eventually get strands representing the string $x\|f(x)\|g(x)$. By making two copies T_1, T_2 of T we can form two solutions: one containing strings of type $x\|f(x)$ and the other containing strings $x\|g(x)$. The problem is combining these two solutions to form the required answer: $x\|f(x)\|g(x)$. We refer to this operation as a *join*.

For technical reasons we formally define the join operations in the following way: Let T_1 be a solution containing strands of the form $\updownarrow x\ y\ \alpha_1$ and T_2 a solution containing strands of the form $\updownarrow \alpha_2\ z\ x$, where α_1, α_2 are fixed 30-mers. That is, all strands in T_1 have an α_1 on their 5' end and all strands in T_2 have an α_2 on their 3' end. The join of T_1 and T_2 is defined as

$$\text{join}(T_1, T_2) = \{\updownarrow \alpha_1\ z\ x\ y\alpha_2 \mid zx \in T_1 \text{ and } xy \in T_2\}$$

None of the biology primitives we have discussed so far enable us to carry out a join of two solutions. We first show how to implement the join operation and then discuss how to apply it when evaluating a group of Xors.

Let T_1, T_2 be two solutions as above. The join of T_1 and T_2 may be carried out as follows:

1. Pour T_1 and T_2 into one container and allow the solutions to mix.

2. Raise the temperature in the container so as to melt the hydrogen bonds and form single stranded DNA.

3. Let the container cool down so that complementary strands will reanneal. We get four types of strands:

   (a) $\updownarrow x\ y\ \alpha_1$

   (b) $\updownarrow \alpha_2\ z\ x$

   (c) $\uparrow \alpha_2\ z\ \updownarrow x\ \downarrow y\ \alpha_1$

   (d) $\downarrow \alpha_2\ z\ \updownarrow x\ \uparrow y\ \alpha_1$

4. Use a polymerase enzyme to complete the sticky ends in the strands. Note that due to the orientation of the strands the polymerase enzyme will only work on strands of the third type.

5. Extract all strands from the resulting solution that contain both α_1 and α_2. This step requires two sequential extracts.

At the present it is not clear whether this procedure works in practice. It is not clear that after we cool the container in step 3 every single stranded DNA will find its mate. The hope is that if initially, the solutions T_1, T_2 contained enough copies of each strand then after cooling, enough strands of type 3 will be formed.

In the spirit of this paper, we will assume for now that the this procedure correctly implements the join operation. We now return to our example of computing two functions $f(x), g(x)$ in parallel. We evaluate the function $f(x)$ in the solution T_1 and $g(x)$ in the solution T_2. For our join operation to work, the strands in T_1 must look like $\updownarrow x \| f(x)$ and the strands in T_2 must look like $\updownarrow g(x) \| x$. Hence, the value of $g(x)$ must be tagged on the left of x as opposed to tagging on the right of x as we have been doing all along. To be more precise, let $x = x_1 \ldots x_n$ be an n bit binary string. Let $f(x) = y_1 \ldots y_r$ and $g(x) = z_1 \ldots z_r$ be r bit strings then the strands in T_1 look like

$$\updownarrow \ S_0 \ B_1(x_1) \ S_1 \ B_2(x_2) \ S_2 \ \ldots \ B_n(x_n) \ S_n \ B_{n+1}(y_1) \ S_{n+1} \ \ldots$$
$$B_{n+r}(y_r) \ S_{n+r}$$

and the strands in T_2 look like

$$\updownarrow S_{n+2r} \ B_{n+2r}(z_r) \ S_{n+2r-1} \ \ldots \ S_{n+r+1} \ B_{n+r+1}(z_1) \ S_0$$
$$B_1(x_1) \ S_1 \ B_2(x_2) \ S_2 \ \ldots \ B_n(x_n) \ S_n$$

Note that we treat the bits of $g(x)$ as though they were in position $n+r$ in the string. The two ends S_{n+r} and S_{n+2r} can be used as the strings α_1 and α_2. When we perform the join operation on the two solutions T_1 and T_2 the resulting strands look like:

$$\updownarrow S_{n+2r} \ B_{n+2r}(z_r) \ S_{n+2r-1} \ \ldots \ S_{n+r+1} \ B_{n+r+1}(z_1) \ S_0$$
$$B_1(x_1) \ S_1 \ B_2(x_2) \ S_2 \ \ldots \ B_n(x_n) \ S_n \ B_{n+1}(y_1)$$
$$S_{n+1} \ \ldots \ B_{n+r}(y_r) \ S_{n+r}$$

The procedure outlined above is a general method for evaluating two functions in parallel on a molecular computer. For the DES circuit there is a natural way of breaking up the work. Recall that one round of the DES circuit consists of the following steps:

1. 48 Xors.

2. 8 table lookups on disjoint groups of 6 bits.

3. 32 Xors.

At the beginning of each round we may break up the work into 8 groups. Each group will evaluate the Xor of 6 bits, perform a table lookup on the resulting 6 bits and finally perform an Xor on the 4 bits resulting from the lookup. We end up with 8 solutions each representing 4 bits of the value coming out of the current round. To combine the 8 solutions we pair them up and do a join to reduce to 4 solutions. We iterate this twice more until we end up with one solution containing the 32 bits coming out of the current round. The entire process is shown in figure 6.

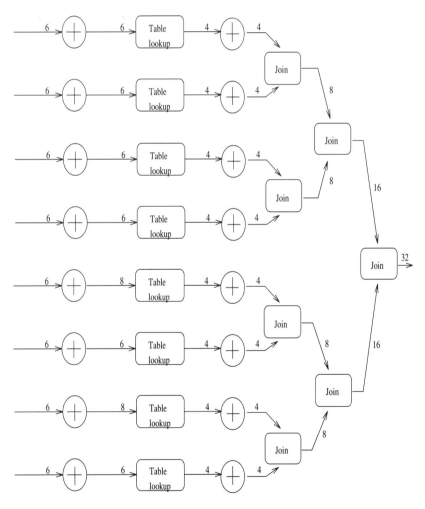

Figure 6: One round of DES circuit

This method of breaking up the work is appealing since it has a number of advantages.

1. Assuming as we did before that it is possible to perform 16 extractions per step, we can evaluate 8 Xor gates at once. This is done by working on the 8 solutions in parallel. Recall that each Xor requires two sequential steps. Thus, we need a total of 292 steps to evaluate the 1168 Xors in the DES circuit.

2. Recall that the table look up operation involved the extraction of strands according to the value of a 6 bit pattern. This step required that we prepare 64 different types of beads. Since now each table lookup is done in a separate solution, we can use the same set of 64 beads for all 8 table lookups in one round. This is a significant reduction in the overhead work for the experiment.

   Note that since each table lookup requires 16 parallel extractions, we can still only perform one table lookup per step. The 128 table lookups in the DES circuit still require 128 steps.

We now need to estimate how many steps it would take to combine the 8 solutions at every round into one solution. There are 3 joins to be made: combine the 8 groups into 4, combine the 4 groups into 2 and combine the 2 groups into 1 (combining the 8 group into 4 requires 4 join operations which can be done in parallel). We saw that each join step requires 2 extractions, i.e. the joins require 6 extraction steps.

Unfortunately there is an additional complication. We explain the problem through an example. Suppose we have 4 solutions T_1, T_2, T_3, T_4 containing strings of type yx, xz, wx, xt respectively. After we do a join on T_1, T_2 and on T_3, T_4 we get two solutions T_{12}, T_{34} containing strings of type yxz and wxt respectively. We would like to do a join on T_{12}, T_{34} to get $zyxtw$, but we can't since the ends of the strings are different(xyz will not join with wxt). This shows that in T_{12} we must copy z to the left to get string of type zyx. Similarly in T_{34} we must copy w to the right to get strings of the form xtw. Doing a join on the resulting solutions will generate the desired $zyxtw$.

This example shows that after we join the 8 groups of solutions to 4 groups we must move some information from one side of the strands to the other. Let T be a solution of strands. Moving 4 bits from on side the strands in T to the other can be done as follows:

1. Separate T into 16 solutions $T_0, \ldots, T_{15}$ according to the 4 bits to be moved. Assuming we are allowed to perform 16 extractions in parallel, this can be done in one step.

2. For all $i = 0, \ldots, 15$ tag T_i with the bit string i on the appropriate side.

3. Set $T \leftarrow T_1 \cup \ldots \cup T_{15}$ and use a restriction enzyme to cut the out the 4 bits that have just been copied.

We have just seen that copying 4 bits requires one extract step and one tag step. Similarly, copying 8 bits would require 2 extract steps and two tag steps.

After we join the 8 solutions into 4 we must perform 4 move operations to prepare the solutions for the next join. Each move involves copying 4 bits and hence takes one step of 16 parallel extractions. Similarly, after we join the 4 solution into 2 we must make another 2 move operations. Each move this time involves copying 8 bits and hence takes 2 extraction steps. This brings us to a total of $4 \times 1 + 2 \times 2 = 8$ steps for rearranging information on the strands.

To summarize, combining the 8 solutions into 1 requires 6 steps for the join operations and 8 steps for the move operations. Thus we need 14 steps per round where each step consists of 16 parallel extractions. Since there are 16 rounds we get a total of 224 steps.

Evaluating Xors takes 292 steps, table lookup take 128 steps and recombination takes 224 steps. Thus, in 664 steps we can evaluate the DES circuit. This figure should be contrasted with the 916 steps required without using the join operation.

8   Summary

The objective of this paper was to show that molecular computers can be used to solve important hard problems. Furthermore the paper provides a concrete example of how a molecular computer might be programmed. We emphasized the fact that boolean gates are evaluated by tagging on their value. Furthermore we saw how to evaluate Xor gates and lookup tables.

The tagging method is inherently sequential since values have to be tagged on one by one. To overcome this problem we introduced the join operation. The importance of this operation is that it enables us to evaluate functions in parallel and then combine the results. Consequently we saw how this improves our ability to evaluate the DES circuit. It should be pointed out that the join operation has never been experimented with in practice. It is currently unknown whether it actually works.

We now briefly summarize our results regarding breaking DES. We showed that assuming one can perform 32 extractions per step, it is possible to break DES in 916 steps. Furthermore, we saw that by using the join operation it is possible to reduce this figure to 664 steps of 16 parallel extractions. We estimate that approximately 10 extraction steps per day is a reasonable figure. Under such an assumption we see that one can break DES in 4 months without using joins and in 3 months using joins.

Bibliography

[1] L. Adleman, "Molecular Computation of Solutions to Combinatorial Problems", Science 266:1021-1024 (Nov. 11) 1994.

[2] D. Beaver, "A Universal Molecular Computer", Penn State University Tech Report CSE-95-001.

[3] D. Boneh, C. Dunworth, R. Lipton, J. Sgall, "On Computational Power of DNA", to appear.

[4] E. Biham, A. Shamir, "Differential Cryptanalysis of the Full 16-round DES", Proceedings of Crypto 1992.

[5] R. Lipton, "Using DNA to solve NP-Complete Problems", Science 268:542-545 (Apr. 28) 1995.

[6] M. Matsui, "The first experimental cryptanalysis of the Data Encryption Standard", Proceedings Crypto 1994, pp. 1–11.

[7] National Bureau of Standards, "Data Encryption Standard", U.S. department of commerce, FIPS, pub. 46, January 1977.

[8] J. Reif, "Parallel Molecular Computation", Proceedings SPAA, 1995, pp. 213–223.

[9] W. Smith, "DNA Computers in Vitro and Vivo", unpublished manuscript.

[10] M. Wiener, "Efficient DES Key Search", Crypto 93 rump session.

Professor Richard J. Lipton

DIMACS Series in Discrete Mathematics
and Theoretical Computer Science
Volume **27**, 1996

Speeding Up Computations via Molecular Biology

Richard J. Lipton[1]
Princeton University
Princeton, NJ 08540
rjl@princeton.edu

Abstract

We show how to extend the recent result of Adleman [1] to use
biological experiments to directly solve any NP problem. We,
then, show how to use this method to speedup a large class of
important problems.

1   Introduction

In a recent breakthrough Adleman [1] showed how to use biological
experiments to solve instances of the famous "travelling salesman
problem" (TSP). Since this problem is known to be NP-complete it
follows that biology can be used to solve any problem from NP. Recall
that all problems in NP can be reduced to any NP-complete one.

However, this does *not* mean that all instances of NP problems can
be solved in a *feasible* sense. Adleman solves the TSP in a totally brute
force way: he designs a biological system that "tries" all possible tours
of the given cities. The speed of any computer, biological or not, is
determined by two factors: (i) how many parallel processes it has; (ii)
how many steps each can perform per unit time. The exciting point
about biology is that the first of these factors can be very large: recall
that a small amount of water contains about 10^{23} molecules. Thus,
biological computations could potentially have vastly more parallelism
that conventional ones.

The second of these factors is very much in the favor of conventional
electronic computers: today a state of the art machine can easily do 100
million instructions per second (MIPS); on the other hand, a biological
machine seems to be limited to just a small fraction of an experiment

---

[1]Supported in part by NSF CCR-9304718. Draft of Dec. 9, 1994.

per second (BEPS). However, the advantage in parallelism is so huge that this advantage of MIPS over BEPS does not seem to be a problem. We will return to this point later on.

Thus, the advantage of biological computers is their huge parallelism. However, even this advantage does not allow any instance of an NP problem to be feasibly solved. The difficulty is that even with 10^{23} parallel computers one cannot try all tours for a problem with 100 cities. The brute force algorithm is simply too inefficient.

The good news is that biological computers *can* solve any TSP of say 70 or less edges. However, a practical isue is that there does not seem to be a great need to solve such TSP's. It appears possible to routinely solve much larger TSP's.

One might be tempted to conclude that this means that biological computations are only a curious footnote to the history of computing. This is incorrect: We will show that it is possible to use biological computations to vastly speed up important computations such as factoring. In particular, we can extend the method of Adleman in an essential way that allows biological computers to potentially radically change the way that we do all computations *not* just TSP's.

Our main first point is that we can extend Adleman[1] to show that we can build a biological computer that can solve any NP problem *directly*. This is important since our biological machines will be limited in the amount of parallelism that they can perform. Thus, solving a SAT problem on 70 variables directly is fundamentally better than using the reduction from SAT to TSP. If one used the standard reduction, then the best SAT problem one could solve in this way would be tiny. Our biological machines will also have some other technical advantages over the original method of [1].

Consider the following computational problem: Given a boolean circuit

$$C(x_1, ..., x_n)$$

of size s; Determine if there is an $x_1, ..., x_n$ so that $C(x_1, ..., x_n)$ is satisfied. Call this problem the (n, s)-*Circuit Satisfaction Problem* (CSP). Our main result is the following:

Theorem 1 *Any (n, n^c) CSP can be solved with at most $O(n^c)$ biological steps.*

Note, the original result from[1]only works for the special case of TSP. There the circuits C are of a very simple form. The main advantage of our result is that the circuit C can be arbitrary.

The second point is that we will show how to use our biological machines as *subroutines* to solve important problems such as factoring. The general idea is the following: suppose that we can solve any CSP of the form $(n, n^{O(1)})$ in constant time. Can we use this oracle to speedup computations?

2   Biological Computations

In this section we will give the exact model of biological computing. We will then show how to prove our main theorem about solving instances of CSP.

The key issue is what is the model of biological computation? In this regard we follow [1] closely. We assume in particular the following simple model. The fundamental concept of a biological computation is that of a set of DNA strands. Since this are usually kept in a test tube will say that a test tube is just a collection of pieces of DNA. Thus, if t is a test tube, from the point of view of a computer scientist, it is just a finite multi-set of strings from {A,C,G,T} . (A multi-set is a set that allows repeated copies of a string.)

The first question is what sets can we start from? The second question is what operations can we perform on test tubes? In our computations we will always start with one fixed test tube: it is the same for all computations. Clearly, this is an advantage over the method in [1]. The set of DNA in this test tube corresponds to the following simple graph G_n. The test tube is formed in the same way that [1] forms the test tube of all paths to find the TSP (see also Appendix). The graph G_n is as follows: It has nodes $a_1, x_1, x_1', a_2, x_2, x_2', ..., a_{n+1}$. Its edges are as follows: there is an edge from a_k to both x_k and x_k'; also there is an edge from x_k to a_{k+1} and from x_k' to a_{k+1}. Then, the paths of length $n+1$ that start at a_1 and end at a_{n+1} are assumed to be in the initial test tube.

The point of this graph is that all paths that start at a_1 and end at a_{n+1} encode a binary n-bit number. For example, the path $a_1 x_1' a_2 x_2 a_3 x_3 a_4$ encodes the binary number 011 in the obvious way.

We need to be able to perform a small collection of operations on test tubes:

1. *Extract*. In this operation we can extract out of a test tube t all those sequences that contain some consecutive subsequence.

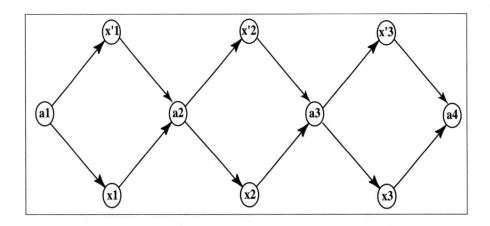

Figure 1: The graph for 3-bit numbers.

2. *Detect.* In this operation we can determine if there is any DNA
   sequence in the test tube at all.

3. *Amplify.* In this operation we can replicate all the sequences from
   the test tube.

We only operate on the DNA sequences from our graph G_n. For these
we use $E(t, i, a)$ to denote all the sequences in t so that the ith bit is
equal to a for $a \in \{0, 1\}$. We, of course, do this by perfoming one
extract operation that checks for the sequence that corresponds to x_i if
$a = 1$ and x_i' if $a = 0$. In the following we assume first that extraction
operates prefectly, i.e. that *all* sequences are extracted. Later on we
will address the issue that extract does not work prefectly.

Before we prove our main Theorem lets prove the following simple
theorem:

Theorem 2 *Any SAT problem on n variables and n^c clauses can be
solved with at most $O(n^c)$ extract steps and one detect step.*

Proof Proof: Let $C_1, ..., C_m$ be the clauses. We will construct a series
of test tubes $t_0, ..., t_m$ so that t_k is the set of n-bit numbers x so that
$C_1(x) = C_2(x) = ... = C_k(x) = 1$. For t_0 use the set t_{all} of all possible
n-bit numbers. Let t_k be constructed; we will show how to construct
t_{k+1}. Let C_{k+1} be the clause

$$v_1 \vee ... \vee v_l$$

where each v_i is a literal or a complement of a literal. For each literal v_i operate as follows: If v_i is equal to x_j, then form $E(t_k, j, 1)$; if it is equal to $\bar{x}_j$, then form $E(t_k, j, 0)$. Put all these together by pouring to form t_{k+1}. Then, do one detect operation on t_m to decide whether or not the clauses are satisfiable or not.

Our next result is the following:

Theorem 3 *Any (n, n^c) FSP can be solved with at most $O(s)$ extract steps and one detect step.*

Here an FSP is just a CSP where the circuit is restricted to be a formula.

Proof   Critical to the proof is the function $TT(f, t)$ defined to be

$$\{x \in t | f(x) = 1\}.$$

Note, f is a boolean function on n bits and t is a set of n bit numbers. Let f be the given formula and let t_{all} be the set of all n bit numbers. Then, we want to construct $TT(F, t_{all})$. Then, one detect operation is enough to find out if there are any x that make f equal to 1. As usual we assume that all internal operations are either $\wedge$ or $\vee$.

Let $C(F)$ denote the cost in steps needed to construct the set $TT(F, t)$ and at the same time also construct $t - TT(F, t)$. We plan to prove by induction on the formula f that $C(F)$ is linear in the size of the formula f.

We now describe how to compute TT. There are four cases:

(1) $TT(x_i, t)$. In this case it is equal to $E(t, i, 1)$.

(2) $TT(\bar{x}_i, t)$. In this case it is equal to $E(t, i, 0)$.

(3) $TT(F_1 \vee F_2, t)$.   In   this   case   form $t_1 = TT(F_1, t)$ and $t_2 = t - TT(F_1, t)$. Then, inductively form $t_3 = TT(F_2, t_2)$ and $t_4 = t_2 - t_3$. The result is then $t_1 \cup t_3$ and the remainder is t_4.

(4) $TT(F_1 \wedge F_2, t)$. In this case first inductively form $t_1 = TT(F_1, t)$ and $t_2 = t - t_1$. Then, inductively form $t_3 = TT(F_2, t_1)$ and $t_4 = t_1 - t_3$. The result is t_3 and the remainder is $t_2 \cup t_4$.

Note, we do the unions by just pouring the two test tubes together. Clearly, it is easy to prove that all the sets are computed correctly.

Moreover, the number of steps to get the formula f is linear in its size.

If one allows amplify steps, then there is a more direct proof of the last theorem. In the induction we only need to operate as follows: Steps (1) and (2) stay the same. For step (3) we just do the "or" by $TT(F_1, t) \cup TT(F_2, t)$. This requires that we copy t, i.e. requires an amplify step. Step (4) is now just $TT(F_1, TT(F_2, t))$.

It may be interesting to note that our method is fundamentally different from usual methods. Logical or's are done by using union's: of course this is not new. However, we do logical and's by a new method. Essentially, a logical and operation is performed by "functional composition". The logical and of F_1 and F_2 is performed by first applying the operator F_1 and then applying F_2.

Theorem 4 *Any (n, n^c) CSP can be solved with at most $O(s)$ extract and amplify steps and one detect step.*

Proof   The key here is just to use the amplify steps to help do the fan out of the circuit.

In all the above theorem's we have assumed that the operations are *prefect*, i.e. that the operations are performed with no error. This definitely needs to be studied. One immediate comment is that the assumption that extract gets *all* of the sequences is *not* needed. If the original test tube has many copies of the desired sequence, then we only need that there is some reasonable probablity that it is correctly extracted to make the theorems work.

3   Using CSP to Speedup Computations

As stated earlier the key question is suppose that one can solve any $(n, n^{O(1)})$ CSP in constant time. How can one use this ability to speedup computations?

Consider any search problem that operates as follows:

```
for each x=1,...,M
if p(x) then
return x
```

where $p(x)$ is some predicate and $M = 2^m$. Assume that p takes polynomial time. Then, the time to do this search is clearly $O(n^c 2^m)$

for some constant c. We claim that the ability to do any $(n, n^{O(1)})$ CSP in constant time allows us to do the above search problem in time

$$O(n^c 2^{m-n}).$$

The proof of this is trivial: just generate $m - n$ of the bits of each x. Then, set up a CSP that is true precisely in the case that some x has the given $m - n$ bits and satisfies the test p. This easily can be set up as a $(n, n^{O(1)})$ CSP. Essentially, we use our oracle to do 2^n tests of p at one time.

Therefore, we can speedup by 2^n any search problem. An important point is that the speedup is actually a bit more complex. Let's compute the exact speedup in the above method. The conventional machine does 2^m tests of p. The biological method uses 2^{m-n} CSP's. Each of these can be done in $O(s)$ biological steps where s is the circuit complexity of p. Thus, the speedup is

$$\frac{2^m s / \alpha}{2^{m-n} s / \beta}$$

where α is the MIPS of the conventional computer and β is the BEPS of the biological computer and α. Clearly, the speedup is $2^n \beta / \alpha$: for reasonable values of n and β and α this is over a trillion!

The critical open question that we cannot yet completely answer is the following: Can we speedup any computation not just the important class of search problems? In any event, it appears that we should attempt to classify those problems which have speedups using CSP's. We expect that there will be progress on this in the near future.

We can, however, say something about some problems that are not exactly search problems. Consider the problem of factoring integers. The inner loop of many fast methods is the search to find $y = x^2 \bmod N$ so that y factors completely over a given "factor basis" (see [2] for details). It is clear that we can speedup the naiive search for such any y by a factor of 2^n. The question however is can we use our methods to speedup the current more sophisticated methods?

Acknowledgements: I would like to thank David Dobkin for a number of helpful conversations about this work. I would also like to thank Len Adleman for taking the time to explain his wonderful construction. I would also like to thank Dan Boneh for his help.

Bibliography

[1] L. Adleman, *Molecular Computation of Solutions to Combinatorial Problems*, Science **266** Nov. 11, 1994.

[2] H. Cohen, *A Course in Computational Algebraic Number Theory*, Springer-Verlag 1993.

Appendix

In this appendix we show how to use Adleman's method to form the initial test tube for our graph G_n. The key to his method is to assign to each vertex a 3' DNA sequence and to also assign another 5' DNA sequence to each edge. Each vertex in G_n gets a sequence of 5' DNA that is of the form $v_i u_i$ where each part is l in length and is randomly selected. The value of l is large enough to avoid accidental matches. Each edge from $i \to j$ gets a 3' DNA that equal to $\hat{u}_i \hat{v}_j$. Here $\hat{x}$ denotes the sequence that is the Watson-Crick complement to x. The initial vertex also adds a 5' sequence that corresponds to its first half; the final vertex also adds its last half. The key is the following: First, every legal path in G_n corresponds to a correctly mathched sequence of vertices and edges. Second, if l is order $\log(n)$, then there will be no other paths. Actually, the latter depends on the random choices, but the chance that it is false can be made as small as one wishes.

DIMACS Series in Discrete Mathematics
and Theoretical Computer Science
Volume **27**, 1996

A DNA and restriction enzyme implementation of Turing Machines

Paul Wilhelm Karl Rothemund
533 South Hudson Apt. #1
Pasadena CA 91101
`pwkr@alumni.caltech.edu.`
WWW: `http://www.ugcs.caltech.edu/˜pwkr/oett.html`

(Turing machines; Universal Turing machines; recombinant DNA; nonpalindromic endonucleases; class IIS nucleases; biological computation; molecular computation; chemical computation, DNA computation)

Abstract

Bacteria employ restriction enzymes to cut or *restrict* DNA at or near specific words in a unique way. Many restriction enzymes cut the two strands of double-stranded DNA at different positions leaving overhangs of single-stranded DNA. Two pieces of DNA may be rejoined or *ligated* if their terminal overhangs are complementary. Using these operations fragments of DNA, or oligonucleotides, may be inserted and deleted from a circular piece of plasmid DNA. We propose an encoding for the transition table of a Turing machine in DNA oligonucleotides and a corresponding series of restrictions and ligations of those oligonucleotides that, when performed on circular DNA encoding an instantaneous description of a Turing machine, simulate the operation of the Turing machine encoded in those oligonucleotides. DNA based Turing machines have been proposed by Charles Bennett but they invoke imaginary enzymes to perform the state–symbol transitions. Our approach differs in that every operation can be performed using commercially available restriction enzymes and ligases.

1   Introduction

The concept of a molecular computer, whose basic operations would be performed chemically instead of electronically, has long intrigued us with the possibility of storing and manipulating information at densities impossible to realize with current computers. For many years, the search for a chemistry rich enough to make a molecular computer has been a focus of our efforts to build them. To many, the chemistry of life with its myriad of enzymes and informationally rich nucleic acids seemed a good candidate—and it has provided us with our first success. Last year, Leonard Adleman used the chemistry of DNA to solve the directed Hamiltonian path problem [1]. This was a very important first step towards realizing molecular computation— similar combinatorial approaches may prove to be the best way to solve NP complete problems. Such an approach does not, however, give us a *programmable* molecular computer that can solve any problem for which we can write a algorithm.

In order to construct a general molecular computer some Universal model of computation (*e.g.* a digital computer, neural network, Turing machine, *etc.*) must be expressed in chemistry. Numerous workers have proposed such translations using theoretical chemistries. Charles Bennett first likened the operation of RNA polymerase to a Turing machine in 1973 [5]. In 1982 [6] he gave a schematic description of a DNA Turing machine using imaginary enzymes capable of recognizing and changing single bases of DNA. Others have since proposed the construction of chemical computers using different sets of hypothetical chemical species and different models of computation. Hjelmfelt *et al.* [14] describe the construction of chemical neural networks which in turn, they proposed, might be used to make other general computers like Turing machines. Models like these that detail how a computation would proceed if we had certain chemicals at our disposal probably must await breakthroughs in protein engineering to be implemented.

Extensions of Adleman's approach ([2], [8]) *can* implement universal models of computation. Adleman [2] presents a series of operation on "DNA test tubes" that simulate a memory model of computation. Boneh *et al.* give a method for simulating nondeterministic boolean circuits. In both of these schemes each reaction performed corresponds to a statement in a computer program or an element of a circuit. This means that the "program" for these computers is actually executed by the chemist.[1] Universal programs can be written for these models but it is suspected that they would be large and require a large number of distinct steps to be implemented.

In this paper we present a method for encoding the transition table of a Turing machine with oligonucleotides, representing a Turing tape, head position, and state, as a single molecule of DNA, and effecting transitions using restriction enzyme chemistry. A total of 6 distinct chemical steps are repeated to simulate a given Turing machine. All of the reagents used in this encoding are commercially available (*New England Biolabs*) and all of the chemical operations are routinely performed by molecular biologists. Using this method a Universal Turing machine (capable of simulating any Turing machine and hence any algorithm) can be constructed.

This paper includes: a review of the Turing machine formalism; a review of the structure of DNA; the operation of restriction enzymes on DNA; a schematic encoding of the three state Busy Beaver Turing machine; the translation of this schematic into real restriction enzymes and oligonucleotides; a description of its extension to a Univeral Turing machine; and a discussion on the practicality of this approach.

2 Turing Machines

2.1 Models of computation

A Turing machine [31] is a *model of computation*—a way of representing and performing a given computation. Turing machines are mathematically equivalent to many other models of computation— cellular automata [16], neural networks [14], and digital computers [13]. Because none of these models of computation (or any other we have found) is more powerful than the Turing machine model we believe that Turing machines embody what we mean when we say

---

[1]It's easy to get a feel for how these models work by simulating them at home: just write out the contents of your computer's memory on little slips of paper and push them around according to a program. The little slips of paper *do* perform a computation, but they hardly evoke the image of a computer that we are familiar with—one of programming a computer, typing `run`, and watching it go. Nevertheless, DNA memory models have a real advantage over paper pushing computers and, perhaps, electronic ones—they allow us to push around 10^{14} little pieces of DNA paper at once! (In [1] 4 x 10^{14} DNA molecules are used to represent edges in a graph.)

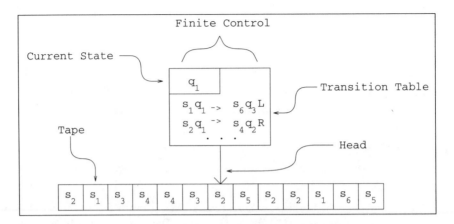

Figure 1: Turing machine model of computation.

something is *computable* (Church's Thesis.) That is, anything for which we can write a procedure, or an *algorithm*, can be computed by a Turing machine. Because of its simplicity and equivalence to other models of computation, Alan Turing's model has allowed us to prove many important results about the nature and limits of computation (*e.g.* the undecidability of the halting problem, and the existence of uncomputable functions.) A correspondence between a DNA computer and Turing's model puts DNA computation on equal footing with *any* other model of computation and allows it to implement *any* algorithm.

2.2 An informal description of Turing machines

A schematic representation of a Turing machine (TM) operating on its tape is given in Figure 1. The tape can be thought of as a sequence of memory *cells* extending indefinitely[2] in both directions. Each cell can store a single symbol from the set $S = \{s_0, s_1, \ldots, s_N\}$. The Turing machine has two parts: a *head*, and a *finite control*. The head points to one cell of the tape and may read a symbol from that cell, write a symbol to that cell, or move right or left to an adjacent cell. At each timestep in the operation of the Turing machine the head performs a compound operation composed of a single read, write, and move. The finite control—the "brain" of the Turing machine—directs the head.

---

[2] *Indefinitely*, in this context, means that if ever our Turing machine runs out of tape we append a few more cells.

A special cell in the finite control contains the *state* of the machine, a member of the set $Q = \{q_0, \ldots, q_P\}$. The rest of the finite control houses the *transition table* which defines how the finite control will instruct the head when the head points to a given symbol s_v and the finite control is in a given state q_x.

For every possible pair (s_v, q_x) the transition table gives a triple (s_w, q_y, m) where s_w is the symbol written by the head, q_y is the new state for the machine, and m is the movement made by the head—L, R, or H, for left, right, and halt, respectively. If m = H the machine enters the a special state q_0, known as the *halting* state, and the computation stops. This allows us to leave q_0 out of the specification of a Turing machine and consider only the non-halting states. Indeed when we state the *size* of a TM we call a machine with j symbols and k *non-halting* states a j x k TM.

Instead of dividing the Turing model of computation into "hardware" and "magnetic media" as we do when we group the head, the transition table, and the state of the finite control together as the Turing machine and place the tape by itself, we might group these elements of the model along different lines: constant elements and variable ones. The variable elements of the model, the head position, the state of the finite control, and the contents of the tape are, taken together, known as an *instantaneous description* or *ID* of a Turing machine. The remaining part of the model, the transition table, is constant and we call it the *machine* part of the model. The Turing machine model, divided in this way, allows us to picture the machine as an *operator* that acts on an ID at time T and produces a new ID at time T+1 [2]. This is the image of a Turing machine that guides the DNA implementation of Turing machines given in this paper.

2.3 An example: Simulation of a 3 state Busy Beaver machine

The Turing machine we show here is a solution to the well known Busy Beaver problem for three state Turing machines.[3] This Busy

---

[3]The Busy Beaver problem for a Turing machine with N states (BB-N) is the problem of designing a N-state Turing machine with two symbols, black and white, that prints the greatest number of black symbols before halting [2]. The example we give is a solution for the three state problem. If we consider the number of black symbols that are printed by the solution to the BB-N problem to be f(N) we can show that f(N) increases faster than any function computable by a Turing

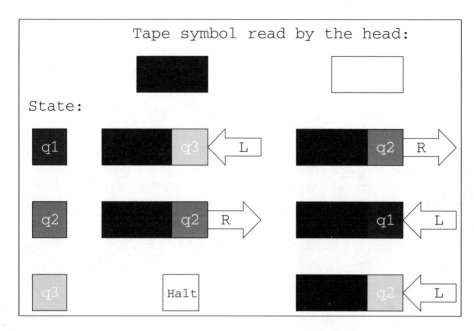

Figure 2: The transition table for a three state Busy Beaver machine.

Beaver machine (BB-3) has a two symbol alphabet S = {B, W} and three (non-halting) states, Q = {q_1,q_2,q_3}. The transition table for the machine is given in cartoon form in Figure 2. The symbols B and W are represented by black and white boxes. Smaller boxes, representing the states q_1, q_2, and q_3 have been assigned three shades of gray. A movement to the left is given by a left arrow and a movement to the right by a right arrow. The next move for the BB-3 machine, if it is in state q_1 and the head points to a W, is (B, q_2, R)—the machine writes a B on the tape, changes to state q_2, and moves to the right. This transition is shown in Figure 3. The operations required to bring the tape from an ID at the last timestep T, to an ID at time T+1 are given at the right of each tape. The machine halts when, if ever, it is in state q_3 and the head points to a B. On a *blank* tape of white symbols it takes 13 steps to print 6 black symbols and halt. While printing 6 black symbols is not a "useful" computation, this BB-3 machine is

machine. f(N) has been determined for N up to N=6 but there is no algorithmic or "mechanical" way to find f(N) for any N. Functions like f(N) are said to be *uncomputable*. This is an example of the results that computer scientists prove with the Turing machine abstraction.

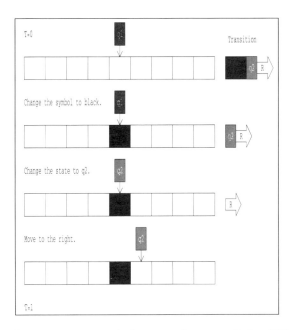

Figure 3: The first timestep of the simulation of the BB-3 on a blank tape.

distinguished because no *smaller* Turing machine can do so.

2.4   Universal Turing machines

While a Turing machine may be constructed to implement any specific algorithm that one can imagine, it would be awkward to have to build a new physical machine every time one wanted to solve a new problem. One's desk would quickly fill up with machines dedicated to different computations. Fortunately, and amazingly, a Turing machine can be constructed that takes as an input a description of another Turing machine and a data tape, and simulates that Turing machine on its own tape. Such a Turing machine is known as a Universal Turing Machine (UTM). The personal computers that we use everyday are good approximations of Universal Turing machines—the programs that they run are descriptions of specific algorithms and hence specific Turing machines. Our personal computers only fall short of UTMs in that their memory cannot be expanded every time we need more storage. The presentation of a molecular computer with the power of a UTM is the goal of this paper.

3   DNA

3.1   DNA, structure and conventions

A single strand of DNA can be likened to a storage tape that can support a four symbol alphabet, S = {A, G, C, T} representing the nucleotides adenine, guanine, cytosine and thymine. These nucleotides are not bonded directly to one another but rather hang from a phosphate and sugar backbone. The DNA strand's backbone has a *polarity;* a sequence of DNA is distinct from its reverse. In order to represent the polarity we name the ends of a piece of DNA according to the structure of its phosphate backbone. One end is called the 5′ end—its terminal phosphate is attached to the 5′ oxygen of a sugar and the other end is called the 3′ end—its terminal phosphate is attached to the 3′ oxygen of the sugar.

Taken as pairs the nucleotides A and T and the nucleotides C and G are said to be complementary. This means that an A-T pair or a C-G pair can form weak, non-covalent bonds known as hydrogen bonds that serve to hold them together. When a stretch of single-stranded DNA encounters another stretch of single-stranded DNA that has a complementary sequence the hydrogen bond interactions between complementary pairs join the two strands in a process called *annealing.* A piece of single-stranded DNA will only anneal to its complement if it has the opposite polarity. When we look at a piece of double-stranded DNA one strand extends from 5′ to 3′ and the other from 3′ to 5′. In this paper we draw double-stranded DNA with the top strand oriented 5′ to 3′ and the bottom strand oriented 3′ to 5′.

3.2   DNA as a computing medium

DNA, to a computer scientist, looks like the tape of a Turing machine. The similarity has prompted others to think of it as a media for computation ([1], [5]). As such DNA has several attractive qualities:

(**A**) *DNA is the genetic material.*

DNA is the storage medium for genes—the plans for the protein molecular machines that perform most of the chemistry in all living things. Our genes determine our morphology, and influence our behaviour. Naturally we are interested in any methods that can modify them in a general way (*e.g.* Turing machines). We also know genes modify themselves with all manner of recombination events. These events may follow rules that allow us to identify them with computation or the generation of languages.

(B) *There are many enzyme-mediated chemical reactions on DNA.*

Nature provides us with a large "toolbox" of enzymes with which to manipulate DNA. These tools range from simple string catenation and string splitting operators like ligases and restriction enzymes to complex copying machinery like polymerases. We steal genes for these enzymes from a variety of organisms, clone them into easy growing bacteria like *E. coli* and harvest them for use in molecular biology.

(C) *DNA is small and easily copied.*

There are 67 atoms per A-T pair and 66 atoms per C-G pair.[4] DNA supports four symbols so this gives a capacity of 1 bit per 33 atoms for double-stranded DNA. The average molecular weight of one base pair is 660 Daltons [20]. This gives an impressive .33 kg DNA / mole bits.[5] In addition, DNA can be amplified very quickly with the polymerase chain reaction (PCR)— over millionfold in an hour [20].

(D) *Reactions between DNA species at equilibrium are completely reversible.*

Charles Bennett has proposed that computers based on DNA or a similar macromolecule would be good candidates for practical reversible computers. Normally computations are carried out in irreversible steps that lose information and dissipate heat. Because enzyme catalyzed operations between DNA species at equilibrium are completely reversible any computations that we embed in them are reversible too. At equilibrium the forward and reverse rates of a chemical reaction are the same. This means that any computation associated with the reaction moves backwards and forwards with equal rates as well; on average no progress is made. To drive the

---

[4]This includes the sugar-phosphate backbone and two Na^+ ions per base pair.

[5]Remember a mole is 6.02×10^{23}!

reaction/computation forward reactants must be added or products removed. The energy dissipated depends only on how fast one pushes the computation to proceed [6]. In principle, one can spend as little energy as desired by slowing down the computation.

4 Restriction Enzyme Operations on DNA

"...the type II nucleases are clearly one of nature's greatest gifts to science."

—Arthur Kornberg in DNA Replication

To create a correspondence between DNA and Turing machines we needed to find some operations on DNA that could be made to correspond with the primitive operations of a Turing machine. We chose to explore the implementation of a DNA based Turing machine with the chemistry of restriction enzymes because we felt it rich and complex enough to do so. Operations on DNA with the most common class II restriction endonucleases proved to have undesirable properties that thwarted efforts to implement a Turing machine. A subgroup of the class II restriction endoncleases, the asymmetric or class IIS restriction enzymes, were found to have operations with properties that lent themselves more readily to Turing machine design.

4.1 Class II restriction endonucleases

Bacteria employ *restriction endonucleases* or *restriction enzymes* to cut double-stranded DNA at or near specific words known as *restriction sites* or *recognition sites*. These enzymes are used to chop up foreign DNA, like that from viruses, which enters the bacterium. The bacterium's own DNA is unaffected because the bacterium's own DNA is chemically modified at the recognition sites in such a way that the restriction enzyme cannot cut it.[6]

Most restriction enzymes recognize 6-8 base pair sequences of double-stranded DNA. These recognition sites have an inverted mirror plane in the middle so that the first half of the site is the reverse of the

---

[6]The chemical tag—a methyl group—that protects the bacterium's own DNA is added to the recognition site by a modification enzyme known as a methylase [22].

complement of the second half of the site. Below the inverted mirror plane is indicated by a |.

```
        5'           3'
 .... ACTG | CAGT ....
        TGAC | GTCA
        3'           5'
```

Biologists call sequences with this kind of symmetry *palindromic*. The cuts made by an enzyme recognizing such a sequence are also made in an inverted mirror symmetric way. One of the most commonly used restriction enzymes of this type is *Eco*R I[7] that cuts at the $\triangledown$[8] in the recognition site below:

```
          ▽
 .... GAATTC ....              .... G         AATTC ....
      CTTAAG          ⟶             CTTAA   +      G
          △
```

Enzymes with this symmetry are known as class[9] II endonucleases or ENases-II of which over 1000 are known [29].

We call the reaction of DNA with a restriction enzyme a *restriction digest*. A restriction digest in which one restriction enzyme is used is known as a *single digest*; if two or three enzymes are employed the reaction is known as a *double digest* or *triple digest* respectively. In practice the treatment of DNA with more than three restriction enzymes at once is rare since different restriction enzymes may cut DNA optimally under different reaction conditions.

4.1.1. Operations of class II restriction enzymes

A single-stranded overhang at the end of DNA cut by a restriction endonuclease can anneal to the single-stranded overhang of a different piece of DNA cut by the same enzyme—for this reason such single-stranded overhangs are referred to as *sticky ends*. When two ends have annealed another enzyme, DNA ligase, may be applied. DNA ligase repairs the cuts in the backbone and a continuous piece of double-stranded DNA is formed. This property of restriction enzyme cut ends allows the following operations on circular DNA molecules known as *plasmids*:

---

[7]We name enzymes for the organisms from which we borrow them—in this case **E**scherichia **co**li.

[8]The positions at which an enzyme cuts are its *cleavage sites* or *cutting sites*.

[9]Biologists use "type" and "class" interchangeably when classifying restriction endonucleases.

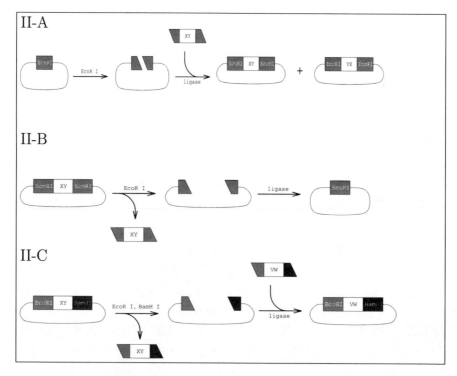

Figure 4: Operations using class II endonucleases.

Operation II-A *Insertion of a DNA fragment into a plasmid without orientation control.*

A single restriction site (*Eco*R I) in a plasmid may be cut,[10] a fragment XY with two matching ends annealed, and the ends ligated (Figure 4, II-A). In this operation all of the overhangs are identical, the fragment is nonorientable, and the reaction yields two products.

Operation II-B *Deletion of a fragment from a plasmid.*

The product of an insertion operation is the substrate or "input" for a deletion operation. A fragment can be deleted if it is flanked by two restriction sites for the same enzyme (Figure 4, II-B). The solution can be diluted until conditions favor circularization and the plasmid DNA can be recycled without the fragment.[11]

---

[10]Here we represent the cut made by Eco RI in a schematic way that emphasizes that the "inverted mirror image" is identical to its mate and hence both sticky ends are identical and self-complementary.

[11]It is interesting to note that this deletion operation is really the same operation as insertion—only the relative concentrations of insert and plasmid DNA differ at the time of ligation.

Operation II-C *Replacement of a fragment in a plasmid with orientation control.*

If a second restriction site, say BamH I, replaces one of the restriction sites normally used in a deletion operation then a double digest using *Eco*R I and BamH I excises an oligonucleotide XY with two different sticky ends. We can then ligate in a fragment VW that has two different ends, assured that its insertion will be oriented (Figure 4, II-C).

4.1.2 Problems with class II restriction enzymes

While the chemistry of class II restriction endonucleases gives us a number of useful operations on DNA strings, several problems[12] are encountered if we try to use them to construct a Turing machine:

Problem 1 *One unique sticky end per restriction enzyme.*

Because the cleavage of a class II Enase occurs within the recognition site each enzyme generates one and only one kind of overhang. Here *kind* is defined by three variables—the length, sequence, and polarity (5′ or 3′) of the overhang. This may seem more of an "obvious property" of class II restriction enzymes than a factor limiting their use but for simple DNA encodings of Turing machines it makes the number of restriction enzymes required unmanageable. If we imagine a DNA encoding of a Turing machine that uses one insertable oligonucleotide to represent each state–symbol transition then we must use one unique sticky end and hence one restriction enzyme for each entry in the transition table. Even the implementation of the smallest UTM known—a 4 symbol, 7 state machine [21]—would require 24 different restriction enzymes and many multiple digests to effect each transition. In the Section 4.1 we will describe restriction enzymes that do not suffer from this "one enzyme-one end" limitation.

---

[12] All of these "problems" are actually just artifacts of the way we choose to limit our use of restriction enzyme chemistry. For example, restriction sites whose subsequences are prefixes for other restriction sites serve as the basis for a whole host of other restriction enzyme operations which can solve all of the problems we list for class II restriction enzymes. The overhangs created by cutting sites with these "partial overlaps" can be chewed up, filled in, and ligated in a variety of ways to destroy the old restriction site and create a new one. Molecular biology, as practiced in lab, is really a collection of just such "clever hacks." To make our restriction enzyme operations independent of the use of specific restriction enzymes we have avoided the use of clever hacks wherever possible.

Problem 2 *Orientability and the problem of Palindromic ends.*

   The identical ends which caused the orientability problem we encountered in Operation II-A above hint at greater problems for symmetric class II Enases. Palindromic sticky ends like those generated by class II Enases are always self-complementary. This means that side reactions, other than those described in Operation II-A, will occur. The large plasmids can, after being cut, ligate together to form larger circular dimers and trimers (Figure 5, A). Fragments meant for

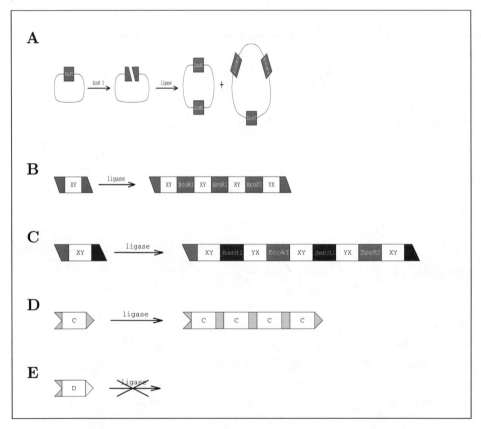

Figure 5: Concatemerization of oligonucleotides.

insertion will ligate to themselves forming multimeric repeats known as *concatemers* (Figure 5, B). Using two restriction sites, as described in Operation II-C, decreases the number of unwanted end matchings in half but will still result in concatemers—the orientation of fragments will just alternate (Figure 5, C). The side reactions resulting from the use of palindromic restriction sites greatly increase the amount purification required between steps and waste reagent oligonucleotides.

Problem 3 *Regeneration of restriction sites*

Because the cleavage sites of Class II Enases occur within their recognition sites the ligation of the overhangs they create results in the regeneration of the original restriction sites. Once a restriction site is used in a plasmid it must be deleted before being used at a different place in that plasmid.

In order to remove an occurrence of a restriction site it must either be flanked by two identical restriction sites as in Operation II-B or be flanked by two different restriction sites as in Operation II-C. These restriction sites are themselves regenerated when the DNA is recircularized and must themselves be flanked by other sites if they are to be removed. This sort of "infinite regression" problem that requires the the use of more and more distinct restriction sites may be solved in several ways but they require more than the chemistry of restriction and ligation.

Problem 4 *Restriction sites "bound" the computation.*

The region in a plasmid that is "accessible" by restriction enzyme chemistry is exactly the region "bounded" by the two most distant restriction sites. While "new tape" can be added at the bounding restriction sites by insertion, each such operation adds a new occurrence of the bounding restriction site—since it is no longer unique it cannot be used as a site of insertion. In order to access a larger amount of DNA, more distinct restriction sites must be added "outside" of the existing ones. This leads to the same sort of infinite regression problem inherent in the deletion of restriction sites makes designing tapes of arbitrary length difficult.

Class IIS restriction endonucleases.

Because class II Enases have recognition sites which are regenerated when fragments are ligated, because they generate palindromic sticky ends which increase side reactions, and because each enzyme can only generate one unique sticky end it seems that they are not good candidates for implementing a Turing machine.

Fortunately, a subgroup of class II restriction enzymes, the class IIS restriction enzymes do not recognize palindromic recognition sites and they cut far away from their restriction sites. One example of a class IIS enzyme is *Fok* I which cuts:

```
                    ▽
  ....GGATGNNNNNNNNNNNNNNNNNN               ....GGATGNNNNNNNNNN            NNNNNN....
      CCTACNNNNNNNNNNNNNNNNN....   ──────▶      CCTACNNNNNNNNNNNNNNNN   +      NN   ....
                   △
```

Where $N \in \{A, C, G, T\}$, independently. These restriction endonucleases are also known as *nonpalindromic* or *asymmetric* restriction enzymes. The asymmetry of class IIS restriction enzymes allows the following new reactions:[13]

Operation IIS-A *Insertion of a DNA fragment with orientation control.*

Since class IIS enzymes cut away from their recognition sites in a stretch of arbitrary DNA one can choose to create an overhang that is non-palindromic. This means that an oligonucleotide C with ends matching the linearized plasmid can be inserted in only one orientation (Figure 6, IIS-A).[14] Here it is easy to see that one restriction enzyme is capable of creating many different sticky ends by varying the arbitrary 4 nucleotide sequence cut by *Fok* I. This solves the "one end-one enzyme" Problem 1 above but concatemers may still be formed (Figure 5, D) and the restriction site persists after the operation.

Operation IIS-B *Deletion of a fragment.*

Simple deletion of a fragment (Figure 6, IIS-B) parallels Operation II-B.

Operation IIS-C *Replacement of an oriented fragment.*

A pair of class IIS Enase sites with opposite directions can be used to prepare two overhangs derived from different sequences (Figure 6, IIS-C.) Now neither the insertion oligonucleotides

---

[13]These and many other operations possible with class IIS Enases are presented in an excellent review article by Szybalski *et al.* [29] One class IIS operation not shown here that may find use in DNA computation is Syzbalski's "universal restriction enzyme", a special adaptor oligonucleotide that recognizes an arbitrary sequence of single-stranded DNA and directs a class IIS enzyme to cut it at an arbitrary position. [30]

[14]In this schematic the recognition site of *Fok* I is represented by the Fok I labeled arrow while its cutting site is represented by the longer "arm." Here the ends created by *Fok* I are drawn in way that emphasizes that the "inverted mirror image" of a given end is not identical to its mate and is not self-complementary.

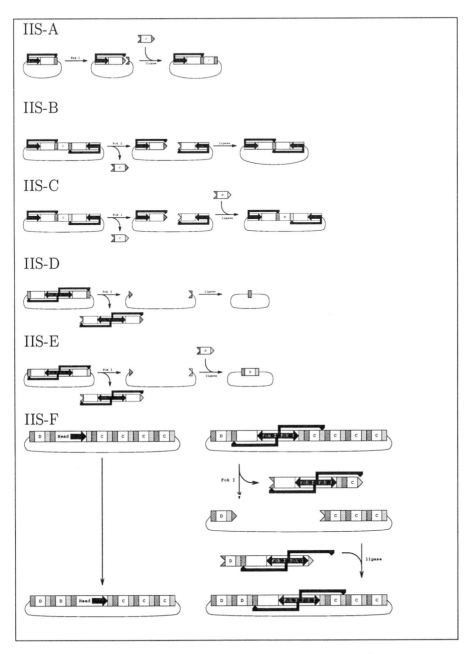

Figure 6: Operations using Class IIs endonucleases.

(Figure 5, E), nor the cut plasmid DNA have self complementary ends and the desired insertion competes with many fewer side reactions. This solves problems Problem 2 but the restriction sites still persist after the operation.

Operation IIS-D *Deletion of a fragment with auto-excision of restriction sites.*

Back to back occurrences of class IIS restriction enzymes (Figure 6, IIS-D) can cut themselves out of a plasmid and the plasmid can be rejoined without the regeneration of a restriction site described in Problem 3.[15]

Operation IIS-E *Replacement of an oriented fragment with the excision of restriction sites.*

This operation is a straightforward combination of strategies used in Operations IIS-C and IIS-D. Here two back to back restriction sites are used to prepare overhangs derived from different cleavage site sequences. This operation solves both Problems 2 and 3.

Operation IIS-F *"Progress"—movement of a sequence through a strand of DNA.*

The left half of Figure 6, IIS-F shows what we mean by "progress"; a sequence Head moves to the right through a string of C sequences and replaces them with D sequences.[16] The right half of the figure demonstrates how two back to back occurrences of *Fok* I could be used to move just such a Head sequence. This operation is really just a variation of Operation IIS-D in which a pair of back to back restriction sites have been added to the right of sequence D in the insert subject to the constraint that the rightmost recognition site has a cleavage site that does not lie entirely in the insert.

This operation highlights the ability of asymmetric restriction endonucleases to "reach out" and cut DNA away from their recognition

---

[15]It is interesting to note that this operation provides the proof that a pair of class IIS restriction sites can simulate *any* class II restriction site. To simulate any class II enzyme f which cuts a restriction site F, all we need to do is replace each occurrence of F with the string FGG'F where GG' represents a pair of back to back restriction sites for a class IIS enzyme g which recognizes G and will cut in the F seqences to create the same kind of overhang that f does when it cuts.

[16]Not unlike the head of a Turing machine!

site making regions of DNA "outside" of the two most distant restriction sites "accessible"—solving Problem 4.

The class IIS restriction endonucleases, then, can be used to solve all of the problems outlined in Section 4.1.[17] In the next section we describe how to use their operations to implement a Turing machine.

5   A DNA Schematic for the BB-3 TM.

To organize the explanation of our DNA schematic we recall one natural division of the elements of a Turing machine: variable elements, and constant ones. First, the variable elements are represented by a single circular DNA molecule; next, the transition table is encoded by oligonucleotide inserts; finally, six chemical steps are described which apply the oligonucleotide encoded transition table to advance the Turing machine one timestep.

5.1   Encoding an instantaneous description

Figure 7 shows the two ways we encode an instantaneous description of the BB-3 TM in which the tape holds the string WBW, the head points at the symbol B, and the machine is in state q1. The representation of symbols, head position, and machine state, as well as the reason we allow two versions of any particular ID, are explained below.

5.1.1 Symbols

Two distinct DNA sequences are used to represent the symbols W and B (Figure 8, A). Each is subdivided into a left and right half.

The DNA Turing tape as shown in Figure 7 is not just the simple catenation of B and W sequences. To each symbol we append, on the left, a short sequence labelled L, and, on the right, a short sequence labelled R (Figure 8, B). Because they are the same in both the B and W symbols these sequences are called the left and right *invariant*

---

[17]Another subgroup of the class II restriction enzymes, those that recognize "interrupted palindromes", have cut sites in arbitrary stretch of DNA between the two halves of what would otherwise be a palindromic site. This means that they do not suffer from Problems 1 and 2 but because their cleavage sites are inside of their recognition sites they cannot solve Problems 3 and 4 as do the class IIS enzymes, without the use of overlapping restriction sites.

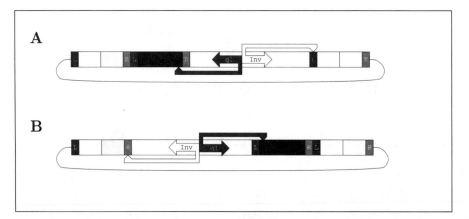

Figure 7: An instantaneous description of a TM in which the tape holds the string WBW, the head points at the B symbol, and the machine is in state q1.

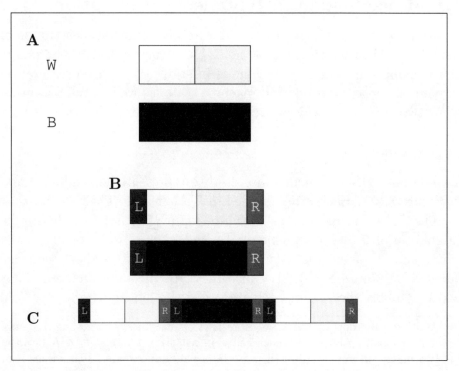

Figure 8: Schematic encoding of the black and white symbols.

sequences. The DNA sequence for a Turing tape (without the head) holding the string WBW is shown in Figure 8, C.

5.1.2 The head

The two back to back asymmetric restriction sites labelled Inv and q1 in Figure 7 represent the head of the Turing machine. The restriction site labelled with the state always points to the current symbol and Inv always points at an adjacent invariant sequence. Our encoding generates two DNA representations for any particular ID because we place the head of the Turing machine "inside" of the tape; one representation (A) positions the head sequence to the right of the current symbol, and the other (B) positions the head sequence to the left. The physical interpretation of these two possibilities is this: if the head is to the right of the current symbol then the last move of the machine was to the left; contrariwise, if the head is to the left of the current symbol then the last move of the machine was to the right.

5.1.3 The state

It is the *spacing* between the recognition site labelled q1 and the current symbol that encodes the state of this Turing machine. Consider the 6 base pair oligonucleotide W', the first half of the symbol W, shown below:

The 4 base pair cutting region of *Fok* I may be used to cut sequence W' in any one of three different cutting *frames* by varying the number of intervening bases between the *Fok* I recognition site and the symbol sequence (Figure 9).

We associate the frame in which a symbol sequence is cut with the concept of state in a Turing machine and call the enzyme used to cut symbol sequences in different frames a state enzyme or state cutter. Each half of a symbol sequence is large enough to be cut in three frames by state cutter enzyme. By carefully picking different DNA sequences for each symbol, the sticky end generated by cutting a given symbol in given reference frame can be made unique. Figure 10 shows the overhangs generated when the symbols W and B are cut in each of the three frames from the right and from the left. The restriction site for

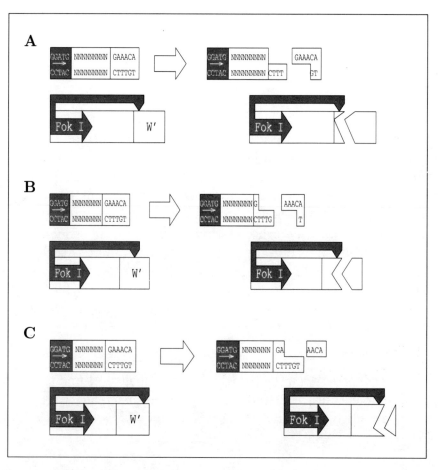

Figure 9: DNA and schematic representations of *Fok* I cutting a six base pair sequence, W', in each of three possible frames. Note that the shape of the cut in the schematic representation differs in each reference frame—the closer the recognition site of *Fok* I to W' the higher the "notch" in the schematic sticky end.

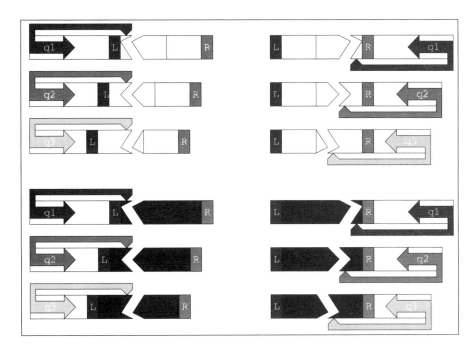

Figure 10: Overhangs generated when black and white symbols are cut by the **state** cutter in 3 different state frames from the left and from the right.

each frame is labelled with the state it represents and shaded to match Figure 2.

5.2 Encoding the transition table

The unique sticky ends which can be generated by cutting a symbol with **state** cutter allow us to ligate, into our DNA tape, a *transition oligonucleotide* which encodes the new state, symbol, and direction of the Turing machine. At any given timestep a Turing machine's last move may have been to the right or left, and its next move may be to the right or left. This means that there are four basic plans for transition oligonucleotides (Figure 11).

Their parts are described below:

Coh is the cohesive end matching the end generated by cutting a given symbol with the **state cutter**.

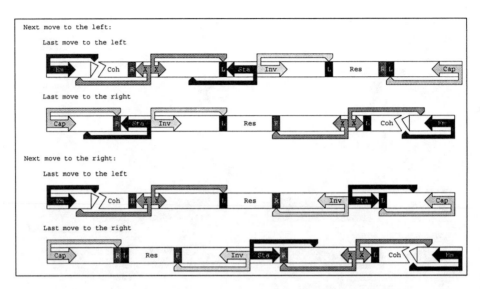

Figure 11: The four forms transition oligonucleotides take. After being cleaved with **end-maker** enzyme to remove the **Em** sequences, each transition oligonucleotide can be divided into 5 sections: **Coh**, the sticky end specific to a particular state-symbol combination, the head sequence (**Inv + Sta**), the result region **Res**, the symbol-excision sites X, and the **Cap** sequence. **Coh** matches a symbol cut during the last timestep by **state** cutter enzyme. If the machine's last move was to the left, **Coh** is the left end of the transition oligonucleotide and vice versa. The symbol-excision sites X are always placed next to the sticky end **Coh**. **Cap** is placed on the opposite end of the oligonucleotide. Between the X and **Cap** sequences we place the head and **Res** sequences. The **Sta** site in the head sequence points in the direction the head sequence will move. The result region is always placed "behind" the head sequence, that is, next to **Inv**.

Sta is a restriction site for the **state** restriction enzyme. It cuts the current symbol according to the state of the machine.

Em is the restriction site for the **end-maker** restriction enzyme. It cuts to create the same size and orientation overhang as the **state** restriction enzyme but it has a distinct recognition site. This site allows the cohesive end to be prepared by cleavage with the **end-maker** enzyme and is used *only* in the manufacture of the transition oligonucleotides.

Res for "result" is a sequence encoding the new symbol.

L and R are two distinct "invariant" DNA sequences that separate symbol pairs in the DNA sequence.

Cap, X, and Inv are class IIS restriction sites whose enzymes all cleave L or R to give the same size and orientation overhang. These sites are used to cleave L or R sequences at various stages of the computation. **Cap** is recognized by the **cap** enzyme, Inv by the **invariant** enzyme, and X by the **symbol-excision** enzyme.

We translate the transition table for the Busy Beaver machine given in Figure 2 using these four forms, making one oligonucleotide with a sticky end Coh to match each end in Figure 10. This generates the the oligonucleotide encoded transition table give in Figure 12.[18]

A j x k TM encoded this way has $2jk$ transition oligonucleotides.

5.3 Making a transition—Going from one ID to the next.

An instantaneous description for BB-3 machine operating on a blank tape is shown in Figure 13.

---

[18]Note that the "whitespace" between a given **state** cutter recognition site and the adjacent invariant sequence R or L differs between an oligonucleotide encoding a change of direction (*e.g.* moving from the left going to the left) or an oligonucleotide encoding a preservation of direction (*e.g.* moving from the left, continuing to the right). If the Turing machine keeps moving in the same direction then there are will always be two invariant sequences between the **state** sequence in the head and the current symbol seqence. If the Turing machine changes direction then for the next move there is only one invariant sequence between them. In our schematic an invariant sequence is the width of 1 cutting frame so the "whitespace" in the transition oligonucleotides has been adjusted accordingly.

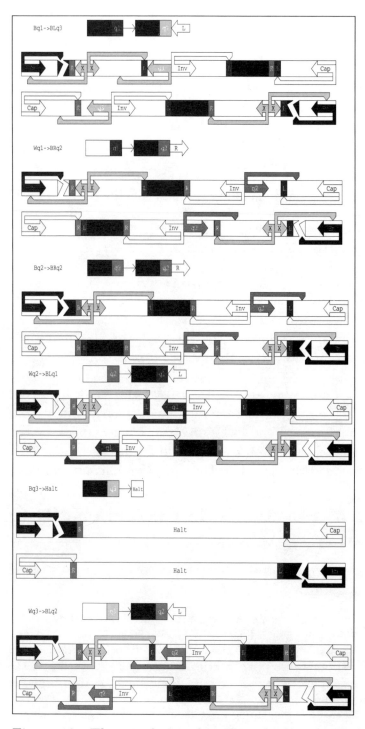

Figure 12: The 12 oligonucleotides which encode the BB-3 machine transition table. Note that in each pair of oligonucleotides the top oligo matches a symbol cut from the left and the bottom oligo matches the same symbol cut from the right.

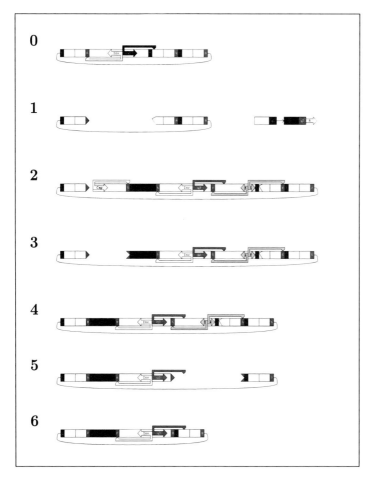

Figure 13: The first step of the BB-3 machine simulated on a blank tape. **0** encodes the first ID. **1** shows the result of cleaving the DNA tape with `state` and `invariant` enzyme. The white cohesive end is specific for the symbol-state combination (W, q1) and matches an oligonucleotide encoding the transition `Wq1 -> Bq2R` shown, in schematic form, to the right of the tape. **2** shows the ligation of the transition oligonucleotide encoding the transition `Wq1 -> Bq2R` to the unique sticky end created in **1**. In **2**, note that the direction of q2's "cutting arm" determines the direction of the head, the spacing between q2 and the invariant sequence R encodes the new state, the black sequence "behind" the head encodes the new symbol and the white sequence to the right of the X recognition site encodes the last symbol. **3** shows the cleavage of the `Cap` sequence protecting the invariant sequence R. **4** shows the intramolecular closure of the DNA tape. **5** shows how treatment with `symbol-excision` enzyme effects the excision of the last symbol. **6** shows the cyclization of the tape to form the next ID.

The series of 6 distinct chemical steps required to take this ID at time T to the ID at time T+1 are shown in Figure 13 and are explained below:

Steps 1-4 replace the head with the correct transition oligonucleotide in an process analogous to Operation IIS-E.

1. Cut the current symbol with the **state** and **invariant** restriction enzymes.[19] The **state** cutter creates an end unique to the current symbol and the current state. The **invariant** cutter cleaves an R sequence to create an end that is the same regardless of what symbols lie to the left of the computation.[20]

2. Mix the twelve transition oligonucleotides in Figure 12 with the DNA Turing tapes.[21] Operation IIS-E assumes that the oligonucleotide insert D is has unique cohesive ends on both ends. Because only the sticky end generated by the **state** cutter is unique we keep the sticky end on the transition oligonucleotide that matches the end generated by the **invariant** cutter protected by the **Cap** sequence. Use DNA ligase to join the transition oligonucleotide to the tape.

3. Cleave the **Cap** sequence that protects the invariant sequence R on the oligonucleotide.[22]

4. Circularize the DNA with DNA ligase. Incorrect ligations can occur if the left cohesive end of an invariant sequence on one DNA tape sticks to the right cohesive end of an invariant sequence on another DNA tape. The reaction can be run at very dilute concentrations of DNA so that intramolecular closure is favored.

Steps 5 and 6 serve to delete the previously read symbol from the tape using the concept of "progress" developed in Operation IIS-F.

---

[19]This may be performed as a double digest or two sequential digests if the **state** and **invariant** enzymes require buffers that are too different.

[20]If the last move had been from the right the **invariant** cutter would have cleaved an L sequence instead.

[21]Remember that these transition oligonucleotides have already been treated with **end-maker** restriction enzyme which cuts at site **Em** to create their unique sticky ends.

[22]Some restriction enzymes have difficulty cutting restriction sites near the ends of oligonucleotides. For this reason a few arbitrary nucleotides (not shown) may have to be added to the end of the **Cap** sequence.

5. Cut with `symbol-excision` restriction enzyme. This cuts away the previous symbol and leaves two matching invariant ends.

6. Recircularize with DNA ligase to join the invariant ends created in 5. As in in 4, reaction conditions should be adjusted to favor intramolecular closure. The `state` cutter restriction site now "points" to the current symbol and the DNA tape once again represents an instantaneous description.

Steps 1-6 are repeated until a `Halt` is incorporated and the computation is done. After any step 6 the computation may checked for the incorporation of a `Halt` seqence. This can be done by PCR amplification of a small aliquot of the computation using `Halt` sequence as a primer. Only halted tapes would be amplified—if detected they could then be sequenced to recover the answer. This scheme is fine if all of the DNA tapes perform the same computation but if different tapes encode different problems some will be lost every time the computation is checked for `Halt` sequences. If a biotin label[23] were incorporated into the `Halt` sequence then streptavidin coated beads could be used to recover halted machines before the PCR amplification step. Only halted tapes would stick to the streptavidin beads so no portion of the unfinished tapes would be removed.

6   Real DNA Sequences for the BB-3 TM.

We now give real sequence and restriction site assignments to the schematic representation presented in Section 5. First, the invariant sequences, `R` and `L`, and the symbols `W` and `B` are assigned in Figure 14.

Next, the restriction sites for the enzymes, *Bbv* I, *Fok* I, *BseR* I, *BsrD* I, and *Bpm* I are assigned to the `end-maker`, `state`, `invariant`, `cap`, and `symbol-excision` sequences (Figure 15).

The unique sticky ends generated by *Fok* I cutting the symbol sequences in each of three different frames are given in Figure 16.

Each overhang is designed so that it is complementary to no other overhang (or complement of an overhang) at more than two positions. This constraint seems to work well in gene construction

---

[23]Biotin is just a functional group that binds strongly to another fuctional group known as streptavidin.

W
TGAAACA GTACA CT
ACTTTGT CATGT GA
B
TGTTAAG GCATG CT
ACAATTC CGTAC GA

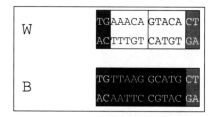

Figure 14: Black and white symbols expressed as real DNA.

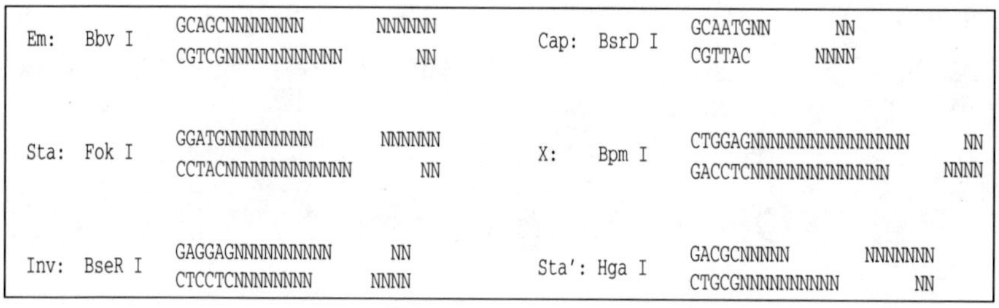

Em: Bbv I
GCAGCNNNNNNNN NNNNNN
CGTCGNNNNNNNNNNNN NN
Sta: Fok I
GGATGNNNNNNNNN NNNNNN
CCTACNNNNNNNNNNNNN NN
Inv: BseR I
GAGGAGNNNNNNNNNN NN
CTCCTCNNNNNNNN NNNN
Cap: BsrD I
GCAATGNN NN
CGTTAC NNNN
X: Bpm I
CTGGAGNNNNNNNNNNNNNNNN NN
GACCTCNNNNNNNNNNNNNN NNNN
Sta': Hga I
GACGCNNNNN NNNNNNN
CTGCGNNNNNNNNNN NN

Figure 15: The restriction enzymes used in a DNA Turing machine, their restriction sites, and the overhangs they generate. `Em`, `Sta`, `X`, `Inv`, and `Cap` are used to implement the BB-3 machine. `Sta'` is an additional enzyme used to implement Minsky's 4 x 7 UTM.

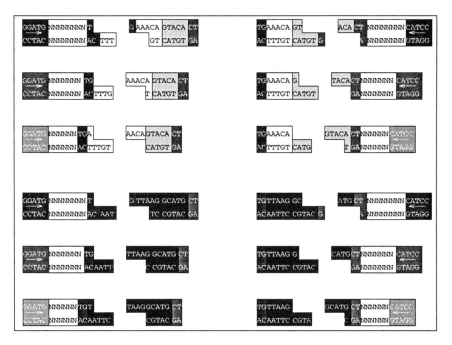

Figure 16: Overhangs generated when black and white symbols are cut by the state cutter *Fok* I in 3 different state frames from the left and from the right. Each overhang is unique and mismatches every other overhang (or complement of an overhang) in at least two positions. Note that the overhangs for state q1 actually include 1 nucleotide of the 2 nucleotide invariant sequences. This is alright—the invariant sequence is regenerated when the correct transition nucleotide is added. This use of part of the invariant sequence as symbol sequence allows us to make the symbols shorter.

	q_1	q_2	q_3	q_4	q_5	q_6	q_7
Y	OL q_1	OL q_1	YL q_3	YL q_4	YR q_5	YR q_6	OR q_7
O	OL q_1	YR q_2	Halt	YR q_5	YL q_3	AL q_3	YR q_6
I	IL q_2	AR q_2	AL q_3	IL q_7	AR q_5	AR q_6	IR q_7
A	IL q_1	YL q_3	IL q_4	IL q_4	IR q_5	IR q_6	OR q_2

Figure 17: The transition table for Minsky's 4 x 7 Universal Turing machine.

[23].   The transition table oligonucleotides are constructed by a direct substitution of the chosen symbol sequences, restriction enzyme sequences and cohesive end sequences into the schematics in Figure 12.[24]

7   Constructing a DNA UTM.

Given enough restriction enzymes, with long enough overhangs and large enough sequences between their recognition and cleavage sites, any Turing machine could be implemented with our technique. In reality, however, the number and variety of class IIS restriction enzymes is limited so only small Turing machines may be constructed. To show that restriction enzyme chemistry is universal, without the design of imaginary enzymes, we demonstrate that our model can implement the smallest UTM reported, a 4 symbol, 7 state machine constructed by Minsky (Figure 17). This machine is constructed using the 4 symbols shown in Figure 18. These symbols have the same property as those chosen for the Busy Beaver machine. Every four base overhang created by a **state** restriction enzyme mismatches ever other such four base overhang in at least two places.[25]   *Fok* I's cutting site can only be shifted through these symbols 4 times to yield 4 different states. To implement the remaining 3 states, an additional enzyme, *Hga* I, is used as **state** cutter. *Hga* I makes 5 nucleotide overhangs (see restriction

---

[24] Actually, an extra 2 nucleotide spacer has to be added between the symbol excision sites X and the sticky end Coh because *Bpm* I excises a longer intervening sequence than is necessary to cut out the last symbol.

[25] The sequences for the symbols W and B are really just the sequences for Y and O less their central two nucleotides.

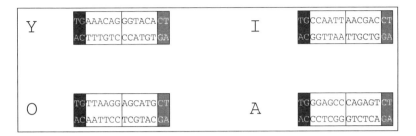

Figure 18: 4 DNA sequences used as the symbols for a UTM.

site `Sta'` in Figure 15) that can be shifted through 3 cutting frames in the given symbols. Because the overhangs *Hga* I creates contain the 4 symbol sequences used in the *Fok* I overhangs for states 1–3 as subsequences they also have the property that any pair of mismatched ends will have at least 2 mismatches. A UTM so constructed has 4*7*2 = 56 transition oligonucleotides. Unfortunately, no counterpart to the **end-maker** *Bbv* I exists for *Hga* I, that is, there is no 5 base overhang cutter that recognizes a different site from *Hga* I for use in preparing the end of those oligos that anneal symbols cut by states 5, 6, and 7. In order to manufacture them we must either synthesize both strands explicitly[26] or synthesize them as two parts, one with a pre-cut *Hga* I restriction site, and the other with an uncut *Hga* I restriction site, that may be ligated together.

8   Discussion

8.1   Error correction

There are many ways for the 6 chemical steps outlined may fail. We have listed the most important error modes and strategies for their minimization below:

(A) *Failed ligations*

---

[26]For the other transition nucleotides which use *Fok* I sites to encode state we need synthesize only one strand and fill in the other with DNA polymerase. Transition oligonucleotides that may be manufactured in this way have the added advantage of being able to be cloned into plasmids or copied by PCR.

Occasionally ligase will fail to join two cohesive ends and will instead leave a piece of linear DNA. This happens during the execution of our DNA Turing machine if a transition oligonucleotide is not incoporated in step 2, or one of the circularizations in step 4 or 6 fails. The "defective" linear tapes which result may be "chewed up" with a DNA exonuclease like Exonuclease III. DNA Exonuclease III only catalyzes the removal of DNA from an open 3′ end so tapes that have closed correctly will be resistant to cleavage. This also means that if ligase manages to make one of the two covalent bonds needed to seal annealed sticky ends in our tape then Exonuclease III can only digest the strand with the nick. Further treatment with a single-stranded nuclease such as S1 will be necessary to degrade the other covalently-closed strand of the damaged tape.

(B) *Incorrect ligations*

DNA ligase can, under certain conditions, ligate mismatched pairs of sticky ends [33]. These incorrect ligations can only occur in our DNA Turing machine during step 2 and may happen in one of two ways. Identical copies of an `invariant` sticky end may be ligated (creating a mismatch at one position) or an incorrect oligonucleotide transition may be joined to the end left by the `state` cutter enzyme (creating mismatches at at least two positions). Tapes with such mismatches can be detected using a single-stranded nuclease such as nuclease S1 [7] which cleaves double-stranded DNA at the single-stranded regions induced by mismatch, or chemical reagents such as hydroxylamine or osmium tetroxide that modify mismatches and make them susceptible to cleavage by piperidine [9]. Once the defective tapes have been identified by linearization, subsequent treatment with Exonuclease III can remove them from the computation.

(C) *Failed restrictions*

Restriction digests are not always complete. After steps 1, 3, and 5 some of the DNA tapes may still carry the head (`Inv` and `Sta`), `Cap`, or `X` restriction sites, respectively. Before the computation can proceed these defective tapes must be removed from the reaction mixture to keep them from interfering with later steps.

Just as a biotin label incorporated into a `Halt` sequence may be used to remove DNA tapes that have finished computing, biotin[27] or

---

[27]Suggested by Nadrian Seeman, personal communication.

other distinct reporter molecules might be used to mark and retrieve tapes from which the head, `Cap`, or `X` sites have failed to be cut. Extra nucleotides can be added at the end of the `Cap` sequence, between the back to back `invariant` and `state` recognition sites, or between the back to back occurrences of the `symbol-excision` recognition sites to carry the reporter groups. After each restriction all DNA tapes which still have the reporter (which should have been cleaved in the last step) may removed using affinity chromatography.

There is no shortage of distinct reporter molecules with which to mark the 3 different functional parts of a transition oligonucleotide. Cholesterol, fluorescein, and dinitrophenyl groups are all available for direct incorporation into oligonucleotides from *Clontech* [10]. The latter two may be retrieved with antibodies. Single-stranded oligonucleotides "side chains" may also be incorporated into the transition oligonucleotides with the use of branched phosphoramidites. Oligonucleotides complementary to these side chains would be used to remove tapes with failed ligations allowing a virtually infinite variety of reporter molecules.

(D) *Incorrect restrictions*

Errors in which a restriction enzyme cuts DNA at a site other than its recognition site are extremely infrequent and have not been well quantified for all restriction enzymes. When restriction enzymes do cut incorrectly it is under non-standard reaction conditions and generally occurs at sequences closely related to their recognition sequence [24]. This means it is good practice to run restriction reactions in their suggested buffers and to minimize the similarity between a particular restriction site and the DNA which surrounds it. Still, this error mode really would have to be explored experimentally under the reaction conditions used for the DNA Turing machine.

(E) *Dimerization of tapes during cyclizations*

In (B) we describe errors in which ligase does the "wrong thing" and joins two unmatched sticky ends. Here ligase does the "right thing" and ligates matched ends, but we choose to interpret it as an error. We have already mentioned the primary means for keeping two different DNA tapes with complementary ends from joining during the recyclization steps 4 and 6—the reactions are run under very dilute conditions so that any molecule of DNA "sees" its own cohesive end more often than

those of others [25]. For similar reasons shorter DNA tapes (as long
as they are not too short and stiff) will also cyclize better than longer
floppier DNA tapes [11]. One design consideration that must be made
for shorter DNA tapes is to ensure that the number of helical turns
in the DNA tape after steps 4 and 6 is integral [27]. This guarantees
that the circularized DNA is unstrained and unsupercoiled. To further
promote intramolecular closure of the DNA tapes we might use any
of a number of DNA binding proteins like catabolite activator protein
(CAP) that are known to bend DNA and promote cyclization [17].

8.2   Solid support

Many systems have been developed for the covalent attachment of DNA
to solid support ([12], [32], [35]). The benefits of performing DNA
chemistry on solid support are twofold. First, the cleanup of successive
chemical steps is greatly simplified. Reagents may be applied and then
simply washed way. The DNA need not be reprecipitated and run on
a gel to separate it from the enzymes oligonucleotides used in the last
step. Second, solid support can effect the separation of different DNA
tapes by maintaining them at a low and constant concentration. This
partially solves the problem of tape dimerization of Section 8.1, (E)
without the need to dilute and reconcentrate the DNA tapes every ID.
Tapes that contain errors are less likely to interact with other correct
tapes so that their removal is not so urgent.

Zhang *et al.* have performed multiple restrictions and ligations
on solid support to synthesize multiply branched DNA molecules. Our
Turing machine implementation requires nothing so complex. A simple
loop encoding the computation attached to the substrate via a single
branched junction would suffice. Figure 19

shows the succesive restriction and ligation of "eyelets" of DNA on
a solid support. Detection of Halt could be accomplished by washing
flourescent or radiolabelled probes for the Halt sequence across the
solid support. Retention of label would indicate that the computation
was done and ready to be cleaved from the solid support.

8.3   Specifications

Very loose estimates for various "hardware specifications" of our DNA
Turing machine are given below:

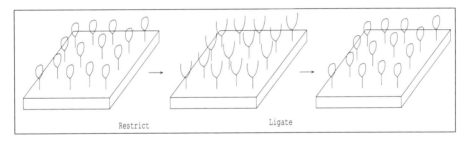

Figure 19: DNA restriction and ligation on solid support.

(A) *Size*

The amount of DNA required for a "useful" computation is ill-defined. Therefore we give only the most basic measures of size for our implementation of Minsky's Turing machine, based on the 16 base pair symbols used to represent its 4 symbol alphabet: atoms—528 atoms/bit; mass—5.28 kg/mol bits; length—2.7 nm/bit; volume—8.5 nm^3/bit.[28] These estimates do not take into account the size or mass of the water or solid support used to maintain this DNA. Restriction digests are normally performed with DNA concentrations around 1 μg DNA / 50 μL water. This means we require about 260 M^3 of water to maintain a mole of bits—or about 1/10th the volume of an olympic sized swimming pool.

How big could an individual DNA tape get? We might want to keep our Turing machines rather small to make the crucial cyclizations in steps 4 and 6 as likely as possible. If, however, we were willing to operate our Turing machines at very low DNA concentrations then much larger tapes could be used. *E. Coli* maintains a 4,700 kilobase circular genome. A DNA tape of this size would have space for 300,000 16 base pair symbols. If, as before, the cells hold members of a 4 symbol alphabet and there are 8 bits per byte this gives us an 80 kilobyte memory per tape.

(B) *Speed*

Restriction reactions require varying amounts of time based on the particular enzyme, temperature, reaction buffer and enzyme concentration used. To a first approximation, the more one spends, the

---

[28]Assuming DNA is a 2 nm wide [28] "cylinder".

more enzyme one can use and the faster the reaction will proceed.[29] Restriction enzymes, however, especially class IIS enzymes[30], are not inexpensive. Further, it is recommended that some enzymes, notably *Fok* I, not be used to *overdigest* DNA. That is, the enzymes shouldn't be used at concentrations much higher than those recommended by the supplier. Even if cost isn't an object this puts one limit on our ability to buy faster reaction times with greater enzyme concentration. We assume, then, that our slowest restrictions are made at "normal" restriction enzyme concentrations so that a complete digest is achieved within an hour. Ligations run at 20°C may be completed in 30 minutes [24]. Assuming that all of the reactions can be performed on solid support so that cleanup is minimal, the 3 restrictions and 3 ligations performed to move from one ID to the next will take on the order of 4.5 hours.

(C) *Energy*

We described, in Section 3.2, Charles Bennett's vision of dissipationless computers. Unfortunately, our machines are not reversible—logically or chemically. The small UTMs we propose to implement do not have reversible transitions and hence lose information. As for the chemistry, restrictions are free, or at least they do not cost us extra energy. The ligations, however, require that one high energy phosphate bond, from the hydrolysis of an ATP molecule to AMP and PP_i, be spent every time a nick is sealed in the phosphate backbone. Two ATPs then, are required to join every pair of sticky ends. Three ligations are performed to move from one ID to the next during which we must spend about 44 Kcal/mole DNA tapes.[31]

Since, in principle, all of the ligation and restriction reactions we perform would be reversible at equilibrium, why don't we take advantage of this and build a near dissipationless computer? For our computer to operate correctly we run it as far from equilibrium as possible! Ligase will perform the "reverse" of its normal operation and nick double-stranded DNA in the presence of AMP [20] but it

---

[29]Using enzymes in a stoichiometric fashion like this is enough to make any chemist shudder.

[30]*Eco*R I is cheap as enzymes go—it costs $50 for an amount that can digest 10 mg of DNA in an hour. The class IIS enzymes *Fok* I and *Bsr*D I cost 10 and 100 times as much, respectively.

[31]Assuming about 7.3 Kcal/mol ATP hydrolyzed [28].

doesn't do so site-specificly. At equilibrium ligase would cut the DNA tapes willy-nilly and our computation would be hopelessly scrambled. For similar reasons we wish to keep the restrictions from running backward—the restriction enzymes have no way of "knowing", for example, which symbols were last excised from which tapes. In a population of tapes at different stages of computation or performing different computations this means that symbols would be randomly reincorporated in to our DNA tapes.

Bennett's computer gets away with operating at equilibrium by performing its transitions as single, site-specific atomic steps; the state, symbol and head position change are all performed at once by a single enzyme. We split our transitions up into steps which, in reverse, are not site-specific and can be mixed and matched to form transition oligonucleotides that were not part of our original Turing machine.

8.4 Flexibility of the model

In Section 7 we demonstrated that Minsky's 4 x 7 Turing machine (and hence all 4 x 7 TMs) could be implemented using commercially availably restriction enzymes. The product of the number of states and symbols (28 in this case) is sometimes taken as a loose measure of the complexity of a Turing machine. Applying this measure to DNA Turing machines we ask: What is the largest state-symbol product that our model can achieve? The answer depends on an number of factors including the specificity and compatibility of enzymes and the number of mismatches we require in incorrectly paired overhangs. Assuming, as before, that we would like to have 2 base mismatches between incorrectly paired overhangs and assuming that we may use all of the class IIS restriction enzymes known in the literature we estimate that our model can simulate Turing machines with a state-symbol product of about 60. Ultimately the size of the Turing machines that can be built with our method will depend on the discovery of new enzymes and our ability to engineer new restriction enzymes with new recognition and cleavage sites. A series of recent papers ([18], [19], [15]) demonstrate that *Fok* I can be mutated to give novel cleavage specificities.

This encoding of Turing machines actually yields a model of computation slightly more powerful than a single tape Turing machine. During the course of a computation the transition table may be changed. This could allow a set of small transition tables to be applied to a DNA tape sequentially, each acting as a subroutine for a larger computation. Additionally, each transition in the transition table need not write a single symbol. The "result" region of each transition oligonucleotide can have an abitrary result string that holds many symbols—or none. This allows us to add cells to the tape if we need more space, or delete cells if this speeds up the computation.

If we choose to run our Turing machines in solution instead of on solid support (Section 8.2) we might decide to take advantage of the plasmid's natural circular boundary condition to implement a different simple model of computation, a TAG system. A TAG system, like a Turing machine uses a 1 dimensional tape but, instead of moving back an forth on a tape, it continuously chews up one end of the tape, and appends strings on the other. If we "circularize" this model of computation and connect both ends of the tape with a read/write head between them we get a model that looks very much like our DNA model, but it moves in only one direction. A TAG system normally reads a symbol, writes a string that is a function of that symbol and then erases some constant number of symbols, P. We can encode this with our DNA model by using P + 1 states. If the head state p is one of the first P states of the machine, then the transition oligonucleotide instructs the head to write no symbol, change to state p + 1 and move to the right. If the head is in state p = P + 1 it writes a result string specific to the current symbol, changes the state p to 1 and moves to the right. Termination in a TAG system can function just as it does in Turing machine with the incorporation of a special Halt sequence into the tape.

8.5   Prospects for molecular computation

Even though our model can perform slightly more complex transitions than a "normal" Turing machine it doesn't seem like our model can ever be much faster than single tape Turing machines—the head still moves one cell at a time. The question remains: Can small Turing machines, compute anything useful in reasonable time? Unfortunately, the smallest Universal Turing machines known are *very* slow. Minsky's machine takes an amount of time exponential on the size of the tape of the machine it is simulating to get from one ID of that machine to another. Stål Aanderaa [1] calculates that his "fairly small" 10 x 6

UTM[32] requires 20,000 steps to advance the simulation of a 2 x 2 Turing by one ID when the simulated machine's tape holds only 6 symbols! Machines that use unary representations of other machines, it seems, are doomed to be slow.

While Universal machines might not be practical Turing machines to implement with DNA, it is possible that smaller special purpose machines which take only very small polynomial time on the size of their input tapes (hopefully linear time!) might perform a useful computation. The dominant DNA computation paradigm (after Adleman and Boneh *et al.*) is to use clever selection techniques to find answers a search space of all possible problems. By reducing the size of the input to these selection based computers we may be able to speed them up. Perhaps small Turing machines could be used as preprocessors for other DNA computers. The real challenge here is to find useful input languages for selection based computers which small Turing machines can generate in a few steps that standard synthetic techniques cannot.

9   Conclusion

In this paper we used the operation we call "Progress" and the concept of cutting frames provided by class IIS restriction endonucleases to propose one way of encoding a Turing machine with DNA chemistry. We assigned real DNA sequences to our schematic that could be used with commercially available enzymes to implement a Turing machine in lab. Finally, we recognized that the Turing machine model, while useful for proving theoretical results, may be too slow to do any practical computation. We do hope that our demonstration that one subset of DNA chemistry can field a Universal Turing machine will motivate others to study the chemistry of biology and come up with a series of operations from which one can build a *practical* universal molecular computer.

Acknowledgements

I would like to thank Max and Judith Rothemund for the generous grant that made my studies at the California Institute of Technology possible. Yaser Abu-Mostafa (dept of electrical engineering California of Technology) provided great enthusiasm, encouragement, and input

---

[32]Aanderaa's machine is too big to be implemented with our encoding using only commercially available enzymes.

for what was an unusual project. Erik Winfree (CNS Caltech) has kept me working and without him, I never would have attended DIMACS. I would like to thank Kai Zinn (dept of Biology California Institute of Technology) for letting me use DNA Strider early on; Greg Fu (dept of chemistry Massachusetts Institute of Technology) and Randy Lee (dept of Chemistry University of Houston) for encouraging my interest in chemistry; Sam Roweis and Len Adleman for their stimulating discussion; Nadrian Seeman for discussion on error correction; and Aditi Dhagat, Barun Chandra, and Amitabh Shah for introducing me to CS.

Finally, I would like to dedicate this paper to the late J.L.A. van de Snepscheut, (Dept of Computer Science California Institute of Technology) for introducing me to the idea of a DNA Turing machine [34]. He would have been excited and pleased by the development of DNA computation.

For complete simulations of the schematic DNA Turing machine presented in this paper, explicit sequences for the BB-3 transition oligonucleotides, or a simulation of the BB-3 machine's first timestep using these oligonucleotides, please browse http://www.ugcs.caltech.edu/ ˜pwkr/oett.html or make a request by email to pwkr@alumni.caltech.edu.

Bibliography

[1] Aanderaa, S., Jervell: A Universal Turing Machine. *Computer Science Logic, Lecture Notes in Computer Science* (1993) 1–4.

[2] Abu-Mostafa, Y.: *Notes on Information and Complexity* (1992) sec. 2.2.

[3] Adleman, L.: Molecular computation of solutions to combinatorial problems, *Science* 266 (1994) 1021–1024.

[4] Adleman, L.M.: On Constructing a Molecular Computer, draft, January 8, 1995.

[5] Bennett, C.H.: Logical Reversibility of Computation, *IBM Journal of Research and Development* 17 (1973) 525–532.

[6] Bennett, C.H.:The Thermodynamics of Computation—a Review, *International Journal of Theoretical Physics* 21 (1982) 905–940.

[7] Burdon, M.G., Lees, J.H.: Double-strand cleavage at a two-base deletion mismatch in a DNA heteroduplex by nuclease S1, *Bioscience Reports* 5 (1985) 627–632.

[8] Boneh, D., Dunworth, C., Lipton, R., Sgall, J.: On Computational Power of DNA, to appear.

[9] Cotton, R.G.H.: Detection of single base changes in nucleic acids, (Review) *Biochemical Journal* 263 (1989) 1–10.

[10] *Clontech Catalog*, Clontech, Palo Alto, CA 1993/1994.

[11] Crothers, D.M., Drak, J., Kahn, J.D., Levene, S.D.: DNA Bending, Flexibility, and Helical Repeat by Cyclization Kinetics, *Methods in Enzymology* 212 (1992) 3–31.

[12] Fahy, E., Davis, G.R., DiMichele, L.J., Ghosh, S.S.: Design and synthesis of polyacrylamide-based oligonucleotide supports for use in nucleic acid diagnostics, *Nucleic Acids Research* 21 (1993) 1819–1826.

[13] Hopcroft, J.E., Ullman, J.D.: *Introduction to Automata Theory, Languages, and Computation*, Addison-Wesley Pub. Co. (1979) sec. 7.6.

[14] Hjelmfelt, A., Weinberger, E.D., Ross, J.: Chemical implementation of neural networks and Turing Machines, *Proc. Natl. Acad. Sci.* 88 (1991) 10983–10987.

[15] Lin, L., Chandrasegaran, S: Alteration of the cleavage distance of *Fok I* restriction endonuclease by insertion mutagenesis, *Proc. Natl. Acad. Sci.* 90 (1993) 2765–2768.

[16] Lindgren, K., Nordahl, M.: Universal Computation in Simple One-Dimensional Cellular Automata, *Complex Systems* 4 (1990) 299-318.

[17] Kahn, J.D., Crothers, D.M.: Protein-induced bending and DNA cyclization, *Proc. Natl. Acad. Sci.* 89 (1992) 6343-6347.

[18] Kim, Y-G., Chandrasegaran, S.: Chimeric restriction endonuclease, *Proc. Natl. Acad. Sci.* 91 (1994) 883–887.

[19] Kim, Y-G., Lin, L., Chandrasegaran, S.: Insertion and Deletion Mutants of *Fok I* Restriction Endonuclease, *Journal of Biological Chemistry* 50 (1994) 31978–31982.

[20] Kornberg,A., Baker, T.A.: *DNA Replication, 2nd Edition*, W. H. Freeman and Company, New York (1993).

[21] Minsky, M.L.: Size and Structure of universal Turing machines using tag systems, *Proc. 5th Symp. in Appl. Math.* American Mathematical Society, Providence, RI, (1962) 229–238.

[22] Naito, T., Kusano, K., Kobayashi, I.: Selfish behaviour of restriction-modification systems, *Science* 267 (1995) 897–899.

[23] Nassal, M.: Total Chemical synthesis of a gene for hepatitis B virus core protein and its functional characterization, *Gene* 66 (1988) 279–294.

[24] *New England Biolabs Catalog*, New England Biolabs, Beverly MA 1995.

[25] Sambrook, K.J., Fritsch, E.F., Maniatis, T.: *Strategies for Cloning in Plasmid Vectors in Molecular Cloning Lab Manual*, Cold Spring Harbor Press (1989).

[26] Siegelmann, H.T.: On the Computational Power of Neural Nets, *Journal of Computer and System Sciences* 50 (1995) 132–150.

[27] Shore, D., Baldwin, R.L.: Energetics of DNA Twisting, *Journal of Molecular Biology* 170 (1983) 957–981.

[28] Stryer, L.: *Biochemistry, 3rd Edition*, W. H. Freeman and Co. New York (1988).

[29] Szybalski, W., Kim, C.K., Hasan, N., Podhajska, A.J.: Class-IIS restriction enzymes - a review, *Gene* 100 (1991) 13–26.

[30] Syzbalski, W.: Universal restriction endonucleases: designing novel
cleavage specificities by combining adapter oligodoxyneucleotide and enzyme moieties, *Gene* 40 (1985) 169–173.

[31] Turing, A.: On computable numbers with an application to the Entscheidungsproblem, *Proc. Math. Soc., series 2* (1936).

[32] Weiner, M.P., Felts, K.A., Simcox, T.G., Braman, J.C.: A method for the site-directed mono- and multi- mutagenesis of double-stranded DNA, *Gene* 126 (1993) 35–41.

[33] Wiaderkiewicz, R., Ruiz-Carillo, A.: Mismatch and blunt to protruding-end joining by DNA ligases, *Nucleic Acid Research* 15 (1987) 7831–7848.

[34] Van de Schnepscheut, J.L.A.: *What Computing is All About*, (1994) sec. 11.5.

[35] Zhang, Y., Seeman, N.C.: A Solid-Support methodology for the Construction of Geometrical Objects from DNA, *Journal of the American Chemical Society* 114 (1992) 2656–2663.

DIMACS Series in Discrete Mathematics
and Theoretical Computer Science
Volume **27**, 1996

DNA computers in vitro and vivo

Warren D. Smith[1]
NEC Research Institute
4 Independence Way
Princeton, NJ 08540
`wds@research.nj.nec.com`

Abstract

We show how DNA molecules and standard lab techniques
may be used to create a nondeterministic Turing machine.
This is the first scheme that shows how to make a universal
computer with DNA. We claim that both our scheme and
previous ones will work, but they probably cannot be scaled
up to be of practical computational importance.

In vivo, many of the limitations on our and previous
computers are much less severe or do not apply. Hence,
lifeforms ought, at least in principle, to be capable of large
Turing universal computations.

The second part of our paper is a loose collection
of biological phenomena that look computational and
mathematical models of computation that look biological.
We observe that cells face some daunting computational
problems, e.g. gene regulation, assembly of complex
structures, and antibody synthesis. We then make
simplified mathematical models of certain biochemical
processes and investigate the computational power of these
models. The view of "biology as a computer programming
problem" that we espouse, may be useful for biologists.
Thus our particular Turing machine construction bears
a remarkable resemblance to recently discovered "RNA
editing" processes. In fact it may be that the RNA editing
machines in *T.Brucei* and other lifeforms are clonable,
extractible and runnable in vitro, in which case one
might get a far better performing Turing machine than all
constructions so far, including our own. The fact that RNA

---

[1]NECI, 4 Independence Way, Princeton NJ 08544

editing *is* a Turing machine may in turn have a lot to do with the origins of life. We also have a computer science explanation for "junk DNA."

KEYWORDS: DNA based computing, nondeterministic Turing machine, RNA editing, transposons, replicon killers, cellular automata, multicellular development, gene regulation, hypotrichous ciliates, junk DNA, biologically based models of computation, NP, PSPACE.

1   Introduction

The main contribution of this paper is to show that DNA and standard biochemical lab techniques may be used to create molecular-size Turing machines in vitro. Turing machines[2] are simple but *universal* computers; that is, you can write a program to run on a Turing machine which will emulate *any* computer performing *any* computational task, such that this emulation will have a slowdown ratio bounded by a *power law*. Because they are DNA molecules, our Turing machines are available in huge numbers, and the programmer can insert *nondeterministic choice* steps into his program which 50% of the molecules[3] will perform one way, the rest the other way. The result[4] is huge parallelism.

The plan of this paper is as follows:

1. We discuss some previous proposals for DNA-based computing.

2. We argue that none of them are competitive with available algorithms on conventional computers.

3. This is at least partially due to the fact that the previous proposals were not universal computers. The second section of this paper shows how to make one.

---

[2]For terms from computer science such as NP, PSPACE, RAM, SAT, Turing Machine (TM), and nondeterministic Turing machine (NDTM), see appendix A, Garey & Johnson 1979, Minsky 1967, or for a popular exposition, Dewdney 1989. For background in biochemistry, see the books in our bibliography, e.g by Lewin, Stryer and Alberts et al.

[3]Actually, any desired percentage is possible. Such steps split the program into two programs operating in parallel from then on.

[4]Our NDTMs also have two capabilities not usually seen in previous discussions: substring insertion capability (cf. §2.2), and an optional and peculiar "random lossy broadcast" interprocessor communication mechanism (cf. footnote 29 in §2.5).

4. Although several other authors have now also devised universal computers based on DNA, our paper, in our opinion, gives the most biochemically realistic discussion. Our quantitative estimates show that the present scheme, and the other authors's schemes *still* look impractical.

5. Because of this, we think that computer scientists who want to work in this area might more productively devote themselves to going in the other direction – i.e. seeing what computer science can say about biology, rather than asking what biochemistry can contribute to computer engineering.

6. So in the third section of this paper, we survey biological phenomena that look computational combined with some speculation about computational models to explain some important biological phenomena. Among these: we have "computer science explanations" for "junk DNA" and "RNA editing in trypanosomes."

7. Last but hardly least, our notion that "RNA editing" is in fact a natural molecular general purpose computational device restores hope that "DNA computing" might be practical, because as we point out, it may be possible to reprogram RNA editing machinery to do whatever we desire. If this works, it will work at continuous speeds of around 1 operation per second, which is thousands of times faster than all the other proposals, including our own.

1.1 Previous work on DNA computers

The notion of using DNA molecules to do computing originated[5] in a paper by Adleman 1994. Adleman showed how to use DNA to solve the "Hamiltonian path problem in directed graphs."

He choose a very simple example, which could be solved by inspection: given a set of 6 "cities" with some interconnecting (directed) "roads," find a path which visits each city exactly once. Adleman synthesized unique short (20-mer) nucleotide "tag" sequences to give each city a "from" and "to" address. Each road was represented

[5]Hjelmfelt et al. 1991 had earlier conceived of the possibility of a chemical Turing machine, but they did not propose any specific chemicals, nor was their abstract published construction complete.

by the 40-mer containing the appropriate city tags in the correct
order. In solution these single-stranded DNA molecules randomly
hybridized to their complements, forming longer strands which were
double-stranded in complementary regions. After allowing ≈ 4 hours
for the strands to hybridize, he used DNA ligase and polymerase to
"sew up" the strands, getting set of DNA molecules representing all
possible paths in the graph. He then performed chemical and physical
operations to extract only the DNA molecules which corresponded to
Hamiltonian paths from this soup[6].

The total number of separation and synthesis steps required grows
only linearly with the size (number of cities and roads) of the problem;
but remember that an exponentially large number of DNA molecules
are required, so this linear bound only works up to a certain point.
Adleman's idea is that since the number of DNA molecules in a soup is
rather large, we get a large constant speedup compared to a sequential
implementation of the same brute force procedure.

Exactly how large this parallelism factor can be made is not clear –
since Adleman only solved a 6 city problem – but it is at least 10^{11} and
no more than 10^{23}. Each individual DNA step took Adleman about a
day, although improvement to under an hour is thinkable. Meanwhile
the comparable steps on a modern digital computer take on the order
of 10^{-8} seconds. Thus, the maximal speedup factor obtainable from
these DNA methods is somewhere between 10^{-2} and 10^{+11}.

Lipton 1994 then showed, beautifully simply, how to use the
same primitive DNA operations as Adleman to solve any "SAT"
problem with N binary inputs and G AND, OR, or NOT gates, in a
number of operations depending linearly on $N + G$. Lipton estimated,
optimistically in our opinion, that it would be feasible to solve such
problems, using his procedure, as large as $N = 70$.

---

[6]Adleman's algorithm can be extended. For example, to find the *shortest*
Hamiltonian path in a graph with small integer edge lengths, use DNA fragments,
corresponding to the directed edges, with appropriate lengths and do a final
electrophoresis step to find the shortest DNAs. Incidentally, we should point out
that there is room for skepticism that Adleman's experiment actually worked as
advertised. The final DNA was not sequenced to exhibit the Hamilitonian path
explicitly; it was merely concluded indirectly that that is the DNA that must have
been produced. It seems to us possible that the initial step of creating "all possible
paths" could in fact have been random annealing amnd random ligating, producing
all possible sequences of edges and cities, whether paths or not. However, that
would not in principle have affected the validity of Adleman's algorithm! – it
would merely have increased the size of its exponential resource requirements.

1.2 Reality: DNA computers aren't competitive

We would now like to express some skepticism. First, ordinary sequential computers are not restricted to using naive brute force algorithms to solve NP-complete problems such as Hamiltonian path and SAT, although the Adleman and Lipton constructions are. Instead, they can use clever "branch and bound" algorithms, which, while still exponential, tend in practice (and[7] in some cases provably) to have much smaller growth constants than the naive algorithm.

The development (Crowder 1980, Padberg & Rinaldi 1987) of programs for rigorous solutions of traveling salesman problems had advanced by 1987 to the point where a 9000-line FORTRAN program on a Cyber 205 solved a real life fully-connected 2392 city TSP in a 27-hour run[8]. Adleman's DNA method, despite its large parallelism factor, couldn't do a 30-city TSP, since $30! \approx 4 \times 10^4 \times N_{\text{Avogadro}}$. "Hard" (Mitchell-Selman-Levesque 1992) random 3CNF SAT problems, with $N = 300$, are solvable by conventional computers (Dubois et al. 1994, Selman et al. 1992) in an average time of 10-20 minutes. This lies rather beyond Lipton's optimistic estimate of $N = 70$ for the capabilities of his DNA based method. The moral is that good algorithms can often buy you speedups much larger than a mere factor of 10^{12} speedup in a naive algorithm[9], so that Adleman and Lipton's DNA methods are not competitive.

Open problem #1: *Can anybody find a class of useful problems*

---

[7]The naive algorithm for SAT takes 2^N steps. Monien and Speckenmeyer 1985 proved that their algorithm for 3-SAT runs in $O(1.62^N)$ steps. An algorithm by Crawford and Auton 1993 solves "hard" 3-CNF instances in time which empirically seems to grow like $2^{N/17}$.

[8]This record (for L_2 distance, points in the plane) is now up to 4461, and the solving time for the 2392-city problem is down to 2.6 hours. TSPs with other distance functions such as $\{1, \infty\}$ and random distances in $[0, 1]$ seem to be much easier.

[9]Nor would the results of Lipton's or Adleman's DNA methods, if they were negative, be a rigorous nonexistence proof, though the probability of error could be made small. Neither Adleman nor Lipton analysed the errors in, nor the physical, engineeering, and rate and thermodynamic constraints on, their processes and how they would behave upon scaling up from toy problems to ones large enough to stymie conventional computers. Actually, this may not matter much, since their exponential growth factors are so bad that they *can't* be scaled up to become competitive. But since *we* have a general purpose computer, we care about scaleup – and the situation looks bad. We'll see that just the ambient rate of DNA damage alone will lead to very severe limitations.

for which some DNA based computing scheme clearly will outperform conventional computers?

Still, molecular computing may have the potential to improve upon the capabilities of conventional sequential electronic computers, and this potentiality is worth investigating.

1.3 Better algorithms wanted - need universal computer

The way in which the good algorithms manage to achieve such good results on SAT and TSP is, extremely crudely speaking, only to branch when they must. Thus when solving SAT, one might prove, by a backtrack search from the current configuration and the use of bounding theorems, that the next three bits must be either a '100' or '110.' In that case, we do not need to consider all 8 possibilities, but only 2 of them, and if this sort of thing were to happen every time, the cost of the search would be $2^{N/3}$ instead of 2^N. The programs that take advantage of such tricks may be thought of as programs on a *nondeterministic Turing machine* but in which the invocations of nondeterminism are carefully controlled and made as infrequent as possible.

We want to be able to duplicate this capability (of allowing clever programming, in which nondeterministic steps are done only at carefully selected places in the program) with DNA, so that we may have our cake (clever algorithms) and eat it too (molecular scale parallelism).

In the present paper, we show how to do this.

1.4 What we do

More precisely, we show that molecular biology lab techniques ought to be[10] useable to create a "nondeterministic Turing machine" (NDTM). (Actually our Turing machine has a finite "tape" hence is not really an NDTM, but this is generally irrelevant in practice, since the tape can be made quite large, as usual. Also, it will later be seen that more tape may be inserted during TM operation, so that in some sense we actually do have an infinite Turing machine.)

---

[10]We emphasize that we have not actually carried the experiment out, unlike Adleman.

For problems in which memory requirements only are polynomially growing functions of the size of the input ("PSPACE"), Turing machines can indeed emulate any other computer with slowdown ratio only polynomial *in the size of the input*. Any program (with nondeterministic statements allowed) for a conventional computer may be executed on an NDTM, and if the (worst case, or average case) time required by the conventional computer is bounded by $P_1(N)A^N$ where N is the number of bits in the input, $A \geq 1$ is a constant, and $P_1(N)$ is a polynomial, then the NDTM will require resources bounded by $P_2(N)A^N$, where P_2 is another polynomial[11]. Thus any clever algorithms which tend to reduce the growth constant A, will still be that way in the DNA version[12]. Also, so long as we are below the point at which we run out of DNA, our NDTM will solve any problem in NP in polynomial time[13].

---

[11]This applies for all the algorithms for TSP and SAT we have mentioned. However, EXPSPACE algorithms, e.g. running in time $O(2^N)$ and consuming 2^N bits of memory, so far as is currently known might require time $\approx 4^N$ to run on a Turing machine.

[12]Assuming they parallelize without much loss. Chakrabarti et al. 1994 and Karp-Zhang 1993 support the notion that backtrack search can be parallelized optimally with high probability. Although these authors are using a computational model which supports interprocessor communication, it turns out that the only sort of communication one needs is precisely the sort of "random broadcast" operation we are capable of supplying (cf. footnote 29), and the "lossiness" there can be overcome, with high probability, by repeated broadcasts.

[13]Some other authors claim to have gone beyond our construction. Our construction shows, crudely speaking, how to solve problems in the class NP with a DNA computer in "polynomial" time. Both Don Beaver (1995) and later John Reif, claimed to have shown how to use a DNA computer to solve problems (again speaking crudely) in the presumably larger class PSPACE. But in fact, Beaver's scheme involves DNA hybridization steps which will require exponentially long annealing times. Reif's scheme avoids that defect, but at the cost of square-rooting the amount of available parallelism; thus, the real class of problems that Reif can solve might better be colloquially described as "$\sqrt{\text{PSPACE}}$." Thus Reif's result does not dominate ours; it is in fact incomparable with it. (The fundamental problem with both schemes is the need to do DNA hybridization steps in soups in which there are exponentially many different kinds of DNA, each one of which is supposed to find its mate.) With the speeds and molecular numbers currently thinkable, this square rooting would cause Reif's scheme to run more slowly than conventional computers.

2   In vitro DNA computer

2.1   DNA NDTM: description and operation

Our Turing machine is a circular loop of partly single and partly double stranded DNA. Actually there are a huge number of such loops, which is essential to give us nondeterministic parallel capability, but let us concentrate on one of them for clarity. Initially, all the TMs are in identical starting states.

Most of the DNA sequence is divided into chunks representing the characters on the Turing machine's "tape." Suppose the chunks are 20 bases long and our Turing machine has an alphabet of size 26. Then we would need to designate 26 special 20-long sequences to represent the letters of the alphabet. There are certain constraints on the choice of these sequences: they need to be dissimilar and they need to avoid long palindromes[14] and certain restriction enzyme subsequences; but these constraints seem to be satisfiable, see appendix B.

Between each such chunk is a separating "comma," which is another special 20-long sequence.

The tape is thus not infinite, but in fact is finite and circular. Since

---

[14]Palindromes, by which we mean a substring S followed after some gap by its reversed complement, can cause hairpins which can stall polymerases and cause bacterial mutation hotspots, (Challberg & Englund 1993). We also wish to avoid translated copies of short substrings to prevent unintentional DNA pairings.

The "modified (k, A) DeBruijn sequences" of appendix B with $A = 4$ are a convenient source of DNA strings which completely avoid mismatches $\geq k$ base pairs long and also completely avoid k-long palindromes. The previous works by Adleman and Lipton could make convenient use of them too. It is not clear how large one can allow k to be before mismatch and palindrome problems become dangerous (annealed DNA is known to mispair, Britten & Kohne 1968; the melting temperature decreases by $\approx 1°C$ per 1% mispairing, according to Wetmur 1976 and Bonner et al. 1973), in either our or the previous works. Even if $k = 5$ is the upper limit, we could still generate $25 \approx 516/20$ tape symbols, which is more than sufficient for Turing completeness (Minsky describes a universal TM with 4 symbols and 7 states; UTMs are now known with smaller (symbol,state) product than the value of 28 for Minsky's machine, for example, Yu. V. Rogozhin found a (6,4) machine, see Robinson 1991.), although not capacious elbow room. Meanwhile, this limitation could put Adleman and Lipton in severe trouble. If Lipton wants to avoid unhappy marriages of length 5, then all C of his L-long DNA chunks must not contain any 5-long substring twice, so $C \cdot (L - 4) \leq 4^5 = 1024$, so with $L = 20$, Lipton would be forced to have $C \leq 64$ and thus he could not solve SAT problems with more than 64 inputs; and the situation is twice as bad if one is also worried about palindromes.

we seriously doubt that anybody will run our NDTM for more than 10^5 steps (and also since, as we will later see, our Turing machine can choose to create more tape) this is not a limitation.

At the far side of the circle (from the head), one would place a special "anchor" subsequence, which could serve to delineate "ends" of the tape and would more importantly be used for anchoring each DNA at a fixed location, rather than allowing it to float freely around in the soup. This anchoring may be accomplished by the incorporation of special "biotinylized bases[15]" into DNA in place of the usual nucleotides by the use of nonspecific polymerases; and then surfaces coated with streptavidin will strongly bind the biotinylized bases[16].

Finally, there is a subinterval of the DNA that is going to represent the single "head" H of the TM and the character L immediately to the left of it.

The scenario is depicted in figure 1. Here "d,o,g," are normal characters on the tape, as are "c,a,t,..." but L, H, and > are special and pertain to the head H and its immediately adjacent tape square L. Actually L is also a normal tape character; the only way in which it is special is the fact that it lies immediately to the left of the head. Note, our Turing machine's head is located *between* two adjacent tape squares rather than on top of one of them. It decides what to do next based on the current head state (here H) and the character (here L) to its left.

7-step process to perform a TM transition

The **first** step is to use a restriction enzyme to cut the DNA at the right of the special subsequence indicated by the symbol '>'. This converts our DNA from circular to linear and leaves "sticky ends." We then remove the restriction enzyme from the container. Since the DNAs are anchored in place, this is simply managed by washing it. At the

---

[15]Discussed in appendix C

[16]See Green 1975, Ruth 1991, and *Methods in Enzymology* volume 184 for reviews of biotin and avidin techniques and properties. (Basics in appendix C.) The fact that the biotin-streptavidin complex, with a heat of formation of 23 Kcal/mole, is very stable to heat (up to 132°C; it also is stable to pH 2-13 and proteases) will be important in our later discussion. One particularly impressive recent technique (Fodor 1993) involving photosensitive endcaps allows one to anchor DNAs to a surface and then add to one end of each bound DNA molecule, a sequence specifying its xy coordinates on the surface, as a binary number!

same time we "melt" the DNA to convert[17] it to single stranded form. (By using hot water, 94-97°C, i.e. well above the melting temperature, which for *E.Coli* DNA is 69°C and for *P.aeruginosa* is 77°C. The non-anchored strands are washed away.)

Restriction enzymes are enzymes which cut double-stranded DNA, often leaving "sticky ends" because they cut the strands at different locations, at specific recognition subsequences. Over 1500 that are known (Roberts 1978, Brown 1991 app. 3) and a few examples are below. Of these, only *Eco*RII and *Bbv*II would at first seem suitable for playing the role of ">," since only they have cut sites at the end of the recognized region. But actually, if we were to adopt the convention that, e.g., every character on the tape begins with A, then one could use *Hind*III, and so forth. (Here A=adenosine, *T*=thymine, *G*=guanine, *C*=cytosine, W={A or T}, N={A,T,G, or C}, and the bottom strand is here written in the $5' \to 3'$ order in which polymerases synthesize DNA. Of the below, only *Hind*III, *Hpa*I, and *Sfi*I were commercially available in 1991.)

```
enzyme name      recognized sequence
                 and cut | locations
                       GGWCC|
EcoRII                 |CCWGG


                 GGANN N|NNTCC
EcoNI            CCTNN|N NNAGG


                 CTTCTGNN NNNN|
BbvII            GAAGACNN|NNNN


                 T TCGA|A
HindIII          A|AGCT T


                 CAA|TTG
HpaI             GTT|AAC
```

---

[17]In fact, fully circular DNA cannot melt, since the two strands are linked. When we say "circular double stranded" DNA we really mean that the second strand does not continue all the way around; the first strand is actually bare in the region near the anchor and only double stranded in the region near the head. This could be assured by the use of bulky artificial biotinylized bases which simply won't pair. See also §2.4.

```
              CC GC|GG
NarI          GG|CG CC


              CCGG|NNNN NCCGG
SfiI          GGCC NNNN|NGGCC
```

The time needed for this step should be between 10 and 100 minutes (Sambrook 5.31, Halford & Johnson 1980).

The *Eco*RI restriction enzyme cuts at its recognition site $(1\text{-}3) \times 10^5$ times more quickly than it makes an erroneous cut at a random site according to Terry et al. 1983, while Halford & Johnson 1980 claimed 5×10^7. Corresponding cut rates on single strand DNA[18] are not known to us. Such imperfect specificity may set upper limits on the size of our Turing machine.

Second (after cooling it down), we pour in a mixture of single stranded DNA from a large container labeled "the transition matrix." This particular mixture is supposed to be available in large quantities and has a specific composition which is independent of the initial state of our Turing machine.

This transition matrix mixture contains some DNA strands which begin with the antisense of , L , H >, continues with a comma ,, and further with the antisense of the *new* L* H* >, subsequence which would result from the TM performing one step, e.g. overwriting L with a new character L* and changing the head state to H*, and ends with the antisense of ,. In the event the TM step called for was *not* overwriting L, but instead moving the head left, the middle part of the transition matrix string would be of the form , H*> , L*, instead of , L , H >.

In the event that a nondeterministic combination of these two operations was called for, then both these strings would be present in the transition matrix. In fact, the "transition matrix" is the collection of all such strings for all possible transitions between all possible old L and H's to new ones, that are allowed by the rules of the Turing machine. However, only the strings which pertain to the particular L and H in our particular TM will bind to it.

This binding will be *both* to the end of the single strand with the old L and H, and also to the other end of the same strand on the other side of the circle with a ,. This makes the DNA circular

---

[18]We will use the fact that restriction enzymes only (except for unintentionally!) cut double strand DNA, later.

once again, after a healing of the nicks by the application of ATP[19] and Ligase. For DNA the size of λ phage (48514 base pairs; 16μm) with sticky ends produced by a restriction enzyme, the circularizing procedure with ligase recommended in the PROMEGA catalog requires 20 minutes (in vivo, λ phage circularizes much more quickly than this); Shore et al. 1981 report that "all linear DNAs tested ranging in size from 242-4361 bp could be completely converted to covalently closed circles in < 1 min at 20°C by sufficient amounts of T4 ligase." The annealing time required (for the transmission matrix strings to bind with their antisense sequences) is harder to estimate, but based on descriptions of similar procedures, kinetics estimates in Britten & Kohne 1968, Wetmur 1968, and Weiss 1968, and especially considering the comparatively short length of these sequences[20], less than an hour should be required.

Because our DNAs are physically separated because they are bound to anchors attached to a rigid object, this relinking to circular form will not accidentally chain several TM DNAs together! Also, note that multiple heads cannot append because the transmission matrix strings are unidirectional – they are not available in reversed order (3' to 5') form!

Third, we use a polymerase and deoxy-nucleotide triphosphates[21] to "fill in the missing pieces" and regain double stranded circular form. (The reader may follow these steps on the figure.)

Stryer page 581 says E.Coli polymerases I,II,III have speeds of 10, 0.5, 150 bases per second. Higher speeds are available by resorting to

---

[19]ATP=Adenosine TriPhosphate, the molecule that serves as the fundamental "currency" of energy in the cell.

[20]Annealing times should grow proportionally to the square root of the effective length L of the DNA strands being annealed and to the reciprocal of the molar concentration: $t_{1/2}^{\text{anneal}} \approx \tau n\sqrt{L}/C$ for L bp long totally unrelated DNA fragments coming in n equinumerous species, where $\tau = 8.9 \times 10^{-6}$mole-sec/liter, and C is the molar concentration of nucleotides. For $C = 100\mu$M and $L = 100$, $n = 100$, this suggests $t_{1/2} < 100$sec. But this could lead to enormous annealing times in at least some instances of scaleups of the schemes by Lipton and by Adleman, e.g. if $n \approx 10^{17}$, $t_{1/2} \approx$the age of the universe...

[21]There is actually no need to "prime" the polymerase for the particular operation required here since all that is *really* needed is to "fill in the gap" on the upper strand in the figure; the lower strand does not really need to be extended as shown. In fact by using 3'-dehydroxylated DNA strands in the transition matrix, if desired, such extensions could be prevented. We should also say that a ligase will be required to seal the nicks left by the polymerase.

the *Taq* polymerase commonly used in PCR[22] but at the cost of a higher error rate. But the replication machinery *in vivo* in E.Coli operates at a much faster rate than that, since generation time is 20 minutes and the DNA is 4×10^6 base pairs for a rate (taking bidirectionality into account, and assuming at most half of the generation time is actually spent replicating) of at least 3333/sec.

The in vitro error rate of the *Taq* polymerase commonly used in PCR is 1/9000 (Kunkel 1989) whereas the Klenow fragment of *E.Coli* gets 1/12000 in vitro. (Nonenzymatic DNA synthesis machines have error and early truncation rates of 1/200 per base, cf. appendix C.) Again, the situation in vivo is much better. Stryer page 635 says that the error rate for T4 phage replication is 1.7×10^{-8} per base per replication. The same figures for *E.Coli* and *Drosophila Melanogaster* (fruit fly) are 4×10^{-10} and 7×10^{-11} respectively[23]. (Nonsubstitution errors are much rarer, unless one intentionally introduces certain intercalating mutagens. Also versions of E.Coli with defective polymerases exist, with much higher mutation rates.)

Fourth, we melt the DNA with a flow of hot water to get back to single stranded circular form.

Fifth, we pour on some more transition matrix strings, except that these strings are slightly different than the ones from step 2 in that they do not extend to the , cut point after the new H* or L*. (In the figure we have called this the "truncated transition matrix.") After annealing we are left with single strand DNA loops but with the portions covered by the old , L , H up to the new H* or L* covered by a second strand.

Sixth, we expose these to a restriction enzyme that cuts at a point well inside the , sequence. Since restriction enzymes only cut double stranded DNA, this only cuts at the commas surrounding the old L and H, which is washed away, and leaves the rest of the DNA unaffected. The "sticky ends" left by the restriction enyme cut are then free to reattach, getting us back to circular but with the old L and H chopped out. Note that in the original restriction in the first step, no such reattachment (which would there, have been undesirable) was possible

---

[22]PCR=Polymerase chain reaction, an important technique for amplifying DNA.

[23]Drake 1991 proposed to explain such numbers, that all DNA based prokaryotes and viruses have an "optimized mutation rate" of ≈ 0.0033 mutations per genome per replication. RNA viruses (Drake 1993) have much higher mutation rates of order 1 per genome per offspring. The DNA replication enzymes alone have much higher error rates than these – of order 10^{-5} per base, and the increase in fidelity above this in vivo is due to additional checking and correcting processes.

due to the immediate melting removing the second strand. Also the present reattachment is made permanent by using ligase to seal the nicks.

Seventh, we use polymerase to get back to double stranded form[24], and we have now completed the TM transition and are ready to start again at step one.

By iterating this 7-step procedure, one may do as many TM transitions as desired. Note that any incomplete reactions will generally either result in a 'no op' or else be corrected/completed next 7-step cycle, so our procedure is fairly robust.

Caveat: For simplicity, we have described a "1-way" Turing machine in which the head can only stay still or move left, and never move right. In fact, it is easy to add some extra bells and whistles of the exact same sort (using different restriction enzymes), to permit the head to move right. We have in fact worked out the details of this, and an 11-step cycle is needed, not a 7-step one.

2.2 Optional extra capabilities – input, output, substring insertion, interprocessor communication

We can make our Turing machine have the capability, not normally seen in theoretical discussions of Turing machines, not only of overwriting a character on the tape, but in fact of *inserting* a substring into the tape, next to the head, in one step.

This leads to interesting possibilities concerning input and output. Suppose the initial tape is totally blank (a periodic sequence blank, comma, blank, comma, actually) except for a head. The head will contain a restriction site of an additional flavor, so that by an single initial insertion of a giant substring (this could be done in vitro with the use of "integrase" and "IHF" the same way λ phage inserts itself into the host genome, or more conventionally with restriction enzyme cuts and ligase) we could write our {input and program} string to all the Turing machines in one initial step. That would take care of input.

The reader may wonder how the initial blank tape is synthesized. It is feasible to synthesize long *periodic* (blank comma blank comma...) DNA with period < 100 base pairs (see Khorana 1968 for the early

---

[24]Actually, it is probably better simply to anneal with the antisense to > , to convert to double stranded form only in one small area.

work in this direction) by synthesizing the periods and linking them with ligases, even though it is infeasible to synthesize a comparably long arbitrary DNA sequence. More importantly, the reader may wonder how the initial program and data tape is synthesized, considering that currently available DNA synthesis machines can only make DNA about 100 base pairs long A slick answer to to both these synthesis problems is to use the Turing machines themselves to synthesize their own input! Specifically, the Turing machine begins by invoking a simple "bootstrap loader" whose purpose is to insert the tape characters forming the input and program (and a handy hoard of blank tape) one (or a few) at a time, using the insertion capability we mentioned earlier. Error correction will be discussed next section.

As for output, in many theoretical discussions of Turing machines it suffices merely to be able to tell whether the machine has entered the "halted" state. For that purpose we can expose our strands to antisense DNA (antisense of 'successfully halted head'), bound to dye. The successfully halted TMs will[25] then be colored. But sometimes one wishes to examine the whole tape. For this purpose one can release all the DNA anchors (photosensitive biotin anchor systems are available which break on exposure to light) and isolate successfully halted TM DNA using the same methods (based on paramagnetic beads bound to antisense DNA for the "halt" state) as were used by Adleman. After amplifying this DNA, e.g. inside bacteria, (by isolating a single bacterium and growing it into a colony, one gets a large number of reliable copies of one particular DNA molecule) one may then sequence it.

2.3 Speed, error rates

For Turing machines with tapes the rough size of λ phage (48514 base pairs), adding up our estimates shows that the 7 step cycle above, which implements a single Turing machine transition, will take a few hours, and will involve an error rate (assuming the steps involving polymerase incur the lion's share of the error) of around 1 incorrect base per 5000 per cycle.

There is reason to believe that if we were to scale up the Turing machine to have a tape N times larger, then the time requirement

---

[25]Another idea: Since we can choose the halt state, we could decide to make it the binding site for some known DNA-binding protein such as the *E.Coli* Lac repressor.

per cycle would increase eventually proportionally to N^p, $1 < p < 1.5$ (with most of the time being required for the circularization steps, see Shore et al. 1981). Also, we would have to use a smaller amount of DNA (more widely spaced apart to prevent chaining in steps 2 and 6) dropping like some inverse power of N. From the standpoint of theory, any such "polynomial slowdown" does not affect the theorems which say that a Turing machine can simulate any other computer with at most polynomial slowdown. But as far as practice is concerned, this is certainly a limitation.

It should be possible to reduce the error rate hugely. In the future better polymerase systems, e.g. with error checking and correcting, may become available for in vitro use. E.g. see Fersht & Knill-Jones 1983. By use of careful procedures and a polymerase from *Pyrococcus furiosus*, Lundberg, Shoemaker et al. 1991 showed that the error rate in PCR could be reduced to 1 nucleotide in 625000, confirmed by Brail et al. 1993.

For the present, we do not need to replicate the entire DNA strand each application of polymerase, it will suffice to terminate the reaction after only the region fairly close to the head H has been converted to double stranded form. Not only will this speed things up, it will also keep the total number of errors low, since errors in replication will only occur in a region within (say) a few hundred bases away from the head. Next, we could use error correcting codes in the transition matrix. We encode tape characters using a radix-4 block error correcting code, and the transition matrix will include corrections for noncodes, so that whenever the head passes over a tape character with a small enough number of substitution errors, it will correct it. As one example of such a code, we mention the quaternary "octacode" (Calderbank et al. 1994) which is a set of 256 words of 8 letters over an alphabet of size 4, such that any two codewords are ≥ 3 mutations apart. To be precise, this code is the linear combinations mod 4 of the rows of

```
11111111
01001213
00103323
00012311.
```

This code will suffice to correct single errors and will allow any choice of 4 bases in any particular 4 places. If we use A=0, C=1, T=2, G=3, then this code also has the property that any two codewords are ≥ 6 *weighted* mutations apart, where mutations between letters only 1 apart mod 4, which are comparatively likely (Stryer p635-7), have weight 1, whereas the remaining mutations $A \leftrightarrow T$ and $G \leftrightarrow C$ have weight 2. Thus it also provides error correction of all weight 2, and detection of weight 3, mutations at no extra charge.

It would also be possible to combine error correcting codes with known DNA-repair enzymes to repair errors throughout our single strand DNA. The method would be to pour in methylated antisense DNA of the tape characters (as before, encoded with radix-4 block error correcting codes) and use ligase to get to double stranded form. Then repair enzymes (Lahue 1989) which take the methylated strands as gospel will snip out erroneous bases on the original single strand and replace them with the right ones, and then a melt will leave us with a corrected single strand once again. Such repair cycles could be undertaken occasionally to repair the entire tape. This scheme actually may be better than the repair methods used by living creatures, and in principle could reduce the substitution error rate near to thermodynamic limits.

But even without any of these repair procedures (which are new, and not standard lab procedures at the moment, hence we are hesitant to rely on them), it should be clear that our methods will suffice to create a Turing machine which will work *for sufficiently small scale problems*. The difficulties all arise when the problems are writ large.

2.4 DNA hydrolysis dissolves your computer & other damage

Speaking of which, another, extremely serious, scaleup problem is that single stranded DNA is less stable to hydrolysis than double strand DNA – and quite a large fraction of the time, we have got single strand DNA sitting around in buffer. Hydrolysis *nicks* in double strand DNA are repairable with ligase, but such nicks in *single strand* DNA are not nicks, but cuts, and there is no way to repair the destruction. Thus, during a long Turing machine run, one may find[26] one's DNA computer dissolving! Our Turing machines could in principle be programmed to

---

[26]Incidentally, the same error rate, hydrolysis and scaleup problems arise in the schemes by Adleman and Lipton, the difference is that we've discussed them. A related problem is that fluid shear commensurate with the pressure of a thumb squirting DNA through a syringe, will tend to cut it into fragments of length $\leq$ 15Kbp (Kornberg p21). DNA in vivo is protected from this by supercoiling and spooling around histones.

try to correct for all sorts of chemical errors occuring locally near the head; but randomly snipping the tape and then washing the head down the drain, is too much!

In practice such hydrolysis problems are usually caused by nucleases introduced by biological contamination and not by the rate of DNA hydrolysis in pure water[27]. According to Lindahl 1993, each human genome (3×10^9 base pairs) in each human cell loses about 2000-10000 purine bases (A & G) to thermal hydrolysis of their N-glycosyl linkages to deoxyribose (depurination), and also about 100 C's spontaneously convert to uracil per day. These rates would be respectively 4 and 150 times higher in single strand DNA. Saul & Ames 1986 also estimate that the human genome suffers 10^5 single strand breaks ("nicks") per day (and tabulate various other damage rate estimates). Fraga et al. 1990 also estimate 10^5 oxidative hits on DNA per day.

In other words, in single strand DNA at 37°C at pH 7.4 we may expect about 10^{-7} errors per base pair per hour plus 1.4×10^{-6} single strand cuts per base per hour.

The long term stability of genetic information in vivo depends on DNA repair enzymes. Since the situation in vitro is probably worse, these may be taken as probable *lower* bounds on the rate of destruction. Shooter & Merrifield 1978 observed that single strand T7 phage DNA was hydrolyzed in a bath of 0.3 molar NaOH and 0.7 molar NaCl at a rate of 1.6×10^{-7} breaks per base per hour at 20°C, and 10 times that rate at 37°C, while Freifelder & Dewitt 1977 found that single strand λ phage DNA rates of $(1\text{-}4) \times 10^{-6}$ breaks per base per hour in various pH media at 75°C. These rates, which arise from intentionally severe conditions, probably may be taken as *upper* bounds (...except when they are smaller than the lower bounds).

Open problem #2: *Can methods be found to reduce the magnitude of this hydrolysis problem?*

Freifelder et al. mentioned that P.F. Davison in unpublished work in $\approx$ 1962 found that DNA in sealed vessels in an N_2 atmosphere had an "extremely low hydrolysis rate," but this "was never followed up." Another possibility – emulating the method used inside our own cells – would be to try to create an almost entirely double stranded DNA computer that somehow uses nickases and/or single strand binding

---

[27]Even if one is extremely careful to avoid contamination from bacteria, minute skin fragments, etc, the enzymes one gets from the supply company are not exactly 100% pure and will contain traces of other proteins, including undesirable ones such as unknown nucleases.

proteins to catalyse the state transitions. However, once one starts building up all sorts of machinery to protect, package, and repair DNA, one soon realizes that what one is trying to create, is also known as life.

For the present, the hydrolysis problem is severe enough to prevent scaleup to computational problems with large (or even moderate!) memory requirements.

2.5   Is this useful? No.

Despite the standard theoretical claims that Turing machines are only polynomially slower than anything else, that polynomial slowdown will probably still be enough to render the scheme suggested here impractical in comparison with the usual sort of computers. When you combine the slow speed with the figures above on the background DNA hydrolysis rate and the limits on restriction enzyme specificities, matters look even worse. (And we have already argued that Adleman's and Lipton's schemes are definitely not of practical interest as computers.) Let's examine this in more detail.

The operations we require for a Turing machine step form a simple 7-step cycle whose repetition could be automated. It seems plausible to us that (we are allowing the possibility of hypothetical future improvements improving the lower bound on the time interval) this cycle could be completed in somewhere between 1 and 200 minutes. With a parallelism factor of 10^{11} to 10^{23}, that would mean that our speedups, as compared to a 100MHz electronic implementation *of a Turing machine*, would be in the range 10^{-1}-10^{+13}. So far, so good, but the speedup ratio compared to a *sensible* electronic computer, on problems for which Turing machines are not especially well suited[28] (i.e., most of them), would be a lot less. Due to the large uncertainties in these numbers (!) we cannot be 100% sure, but we suspect that for most computations, little or no advantage will be obtainable.

Bottom line – a Revealing Calculation: Assume $10\mu g$ of 1000-bp loop DNA, i.e. 10^{13} molecules, each $.3\mu m$ in circumference, bound to some large (spongelike?) surface, and that each 7-step cycle takes 1hr. Then 1000 NDTM steps on tapes with ≈ 30 characters could be accomplished in about 1 month, after which time computations would

---

[28] As a rare example of a task to which TMs are especially suited, consider the task of computing (or trying to) the fifth "busy beaver number" (Brady 1983, Dewdney 1985, Buro 1990, Michel 1993).

have to cease since a good fraction of the DNAs would have hydrolyzed. Meanwhile a 100MHz electronic TM would only have accomplished 2.5×10^{14} steps, i.e. 40 times fewer steps, *but* we doubt this factor of 40 is enough to overcome the handicaps of being a small Turing machine, and if you'd really cared you might have built 40 electronic TMs. □

However, we *may* be able to use a very large transition matrix, much larger than the matrices found in most theoretical discussions of Turing machines. This might make the task of programming the machine less onerous and involving less steps to do something interesting, than anyone (unhappily) familiar with Turing machines, might at first think. It would also allow the use of error correcting codes as discussed before. (Transition matrices with N times more elements will require N times more annealing time.)

One could also return Turing machines which have reached an unsuccessful halt state to the computational pool, by the crude method of programming such machines to back up along the nondeterministic computational tree, until they are ready to resume computing (in other words, not all backtracking has to be done using nondeterminism). Less crude methods are unavailable to us since we lack good interprocessor communication mechanisms[29].

Open problem #3: *What is really wanted is not a nondeterministic Turing machine, but rather, a nondeterministic RAM or PRAM. Can anybody show how to construct one with DNA?*[30]

Incidentally, although we argued early that "the maximal speedup factor obtainable from these DNA methods is somewhere between 10^{-2} and 10^{+13}" because the maximal parallelism is between 10^{11} and 10^{23}, we point out that methods might be conceived in which computations happen at more than one place on *each* DNA molecule, e.g. a multihead Turing machine or 1D cellular automaton. If that were managed, an additional factor of perhaps 10^2-10^6 would be exploitable. We do not know how to accomplish this using standard lab techniques,

---

[29] Actually by the use of λ phage Integrase and IHF (integration host factor) enzymes, and/or the excision factor Xis (Ptashne 1992), one would be able to snip out and circularize substrings from DNA, and later reinsert them at in the DNA of other (random!) Turing machines. Integrase always does this insertion and excision at a specific 15 base pair sequence on the host (and message) DNA (Mathews & van Holde page 883). This would allow a rather strange "lossy random broadcast" communication mechanism.

[30] Without needing to resort to "biosteps" which actually require exponential time, and while still having a very large usable memory?

but certainly this would be relevant for modeling the computational capabilities of DNA in vivo.

2.6   Summary – our machine and previous ones and their problems

Many previous schemes live in an imaginary world in which all reactions have 100% yield, DNA never gets damaged, and all steps happen quickly, etc. For the first time in the present paper we have analysed the reality of the situation. Although our scheme is theoretically more powerful than some previous schemes which were not universal computers, it also does not suffer from some severe practical problems that afflicted earlier schemes. Despite this, our scheme *still* does not appear practical.

1. Ours is the first scheme that shows how to make a *universal computer* with DNA.

2. We also suspect that our scheme ought to be less prone to error and simpler to do than the previous schemes of Adleman and Lipton.

To dwell on the latter item for a moment: We avoid PCR, and use only 1 step of polymerase, thus allowing the convenient use of a high reliability polymerase. We do not need to duplicate the whole genome and do not need to rely on the constant use of polystyrene antisense bead and amplify steps, which sound like a big source of possible error and exponentially multiplying losses, and in addition were the biggest percentage of the 7 day lab time requirement for Adleman. We do not require unrealistic annealing of long DNA sequences in which each DNA strand has to pick out its mate from 10^{17} wrong alternatives in the soup. We've demonstrated the capability for error correction. We do not require pre-synthesis of unrealistically long DNAmers. Our recommendation of modified DeBruijn sequences (appendix B) for Turing machine symbols avoids or reduces problems with hairpins and mismatches that would have plagued earlier schemes. Our entire cyclic computational process takes place for DNA bound to a single substrate, and thus presumably immune to loss, in a single receptacle, an important practical convenience advantage.

However, we are still susceptible to unrepairable DNA damage due to restriction enzyme cuts at noncognate sites, contamination (e.g. by stray nucleases; such contamination may be unavoidable), and hydrolysis damage[31]. This situation is exacerbated by the excruciatingly slow speed of the computations we run.

3 Biological significance?

Does the fact that a Turing machine can be made out of DNA have any biological significance? Yes!! Certainly repressor and promotor protein feedback loops (cf. §3.6), intron excision, transposon, genetic switch, and antibody synthesis mechanisms (to name a few) ought to give cells considerable computational power. And it is also obvious that in fact cells do use such power to make the little logical decisions crucial to maintaining homeostasis, as well as pivotal growth decisions, constantly. Our particular scheme is very crude in comparison; indeed, antibody synthesis is a computational search task on which cells apparently outperform the best computer hardware and software available today.

But for some reason, this point of view is not very prevalent[32]. We are now going to try to change this. We would like to inspire the new and sometimes useful paradigm of "thinking of biology as a computer programming problem."

First, we are going to show that our Turing machine construction is astoundingly similar to "RNA editing" processes that have been discovered during the last decade, which in turn seem similar to functionalities required by the hypothetical early "RNA life." So this may cast light both on the questions of why RNA editing exists, and also on the question of how life began. Second, we will show that cells face some severe computational problems. Third, we will survey some known biochemical mechanisms inside cells which seem to have computational capabilities, and discuss what those capabilities are.

---

[31]Incidentally, we don't suffer exponentially from at least *some* reaction yields being below 100% because we'll simply be re-running uncompleted reactions during future 7-step Turing cycles, but any parasitic reactionm such as hydrolysis, removing some tiny percentage of our DNA each step will cause exponentially multiplying losses during long computations. Remember that typical computations involve huge numbers of steps

[32]For example, a quote from the otherwise excellent book (Alberts et al. 1994): "...it is hard to imagine how a cell could keep a long-term account of its division cycles and halt after completing 50." A Turing machine could easily handle this problem, emulate any finite state machine, etc.

This is a loosely organized collection of different biological phenomena that look computational, together with some computational models which look biological, and a few new biological hypotheses suggested by our viewpoint. It is hoped that its informational and inspirational qualities will make up for whatever lack of coherence and speculative character it displays.

3.1   RNA editing in Trypanosomes

This is a long story, but we cannot resist recounting it.

The trypanosome *T.Brucei*, which causes African sleeping sickness, has many unusual properties. It is a single cell, $\approx 4\mu$m long, with a flagellum. Its outer wall is covered with glycoproteins which change, apparently randomly, every few days, thus keeping a step ahead of the immune system of the host. Also in the cell are a nucleus, and a single long tubular mitochondrion which contains a so called "kinetoplast" body.

Inside the mitochondrion is DNA, which comes in both "maxicircles" (≈ 30Kbp) and "minicircles" (400-2500bp each; 300-400 kinds, says Correll et al. 1993). The minicircles are interlinked like chain mail. The maxicircles encode at least 13 genes for mitochondrial proteins, but some genes seemed to be mysteriously missing, a large portion of the genome seemed to be noncoding, and some genes seemed to be in the wrong reading frame – very mysterious. Also, the minicircles apparently don't code for any proteins and so their function was initially mysterious too.

These mysteries were resolved by the exciting discovery (Benne et al. 1986) of "RNA editing." According to Maslov 1994, "RNA editing occurs in all trypanosomatid species so far examined," but *T.Brucei* is the most studied one.

Many of the genes in *T.Brucei* maxicircles are in a so called "encypted" form. They are transcribed into messenger RNA (mRNA) as usual. But then this messenger RNA is "edited" over 100 times. For example, the cytochrome C oxidase gene transcript, which is 712 bases long, is edited by the insertion of 398 U's at 158 sites, and the deletion of 19 U's from 9 sites. This editing changes the reading frame with abandon and certainly totally changes the message. At its conclusion, the messenger RNA contains the "plaintext" for cytochrome C oxidase, and is used for protein synthesis. The RNA editing in trypanosomes that has been found so far consists entirely of site specific insertion and deletion of U's, but other organisms have been found which do other kinds of RNA editing[33]. Mutant trypanosomes without RNA editing, die (Stuart 1991).

The editing in *T.Brucei* is performed by "editosome" protein complexes (Göringer et al. 1994, Seiwert & Stuart 1994) with the aid of "guide RNAs." The guide RNAs, which are 35-78 bases long, contain an "anchor" portion 4-14 bases long antisense to the (apparently arbitrary) pattern that they recognize, plus a additional subsequence defining how many (up to 32) U's are supposed to be inserted (or deleted) from (up to 10) sites; they also contain various hairpinned structural portions and a scorpionlike tail containing U's which can be donated to the edited RNA. The gRNA tails can later be recharged with U's by "TUTase" (Simpson 1990) and UTP. The gRNAs will form duplexes with mRNA without requiring protein in some cases.

The guide RNAs are encoded in the minicircles (Benne 1992) and also in intergenic regions of maxicircles.

According to Simpson 1990 "It remains to be seen if the trypanosome type of RNA editing is present in higher eukaryotes." In most genes known so far in most eukaryotes, mammals included, the only RNA editing which goes on (but it happens extensively!) is the excision of introns (substrings) by "spliceosomes." What makes *T.Brucei* and its kinetoplastid kin different is that more than simply splicing out introns is going on – there are insertions and deletions – and also each of these editing operations is apparently coded for by a site-specific and edit-specific "guide RNA," which apparently, in principle, could have specified anything. There appears to be one guide RNA for each edit operation on each cryptogene. Meanwhile in mammals, so far as is known, there are only a few kinds of spliceosomes and although they locate (somewhat) specific sequences, these sequences seem to be

---

[33]Examples: The slime mold *Physarum polycephalum* does 54 C-insertion edits on at least one mitochondrial gene. Two unencoded G residues are found in mRNA from paramyxovirus SV5. Maize (Kumar & Levings 1993) edits its C-atp6 gene by performing 19 specific C to U alterations. U→C changes have also been observed in plant mitochondria, while U→A, U→G, and A→G single nucleotide conversions are observed in *Acantha Amoeba Castellanii* mitochondria (Lonergan & Gray 1993). Mammalian apoLipoprotein B comes from an mRNA in which a single C→A, conversion was performed, but this is known (Greeve et al. 1991) to occur via an entirely different sort of mechanism from the one in *T.Brucei*; meanwhile the RNA editing in yeast (*Saccharomyces Cerevisiae*) seems to be more similar to *T.Brucei* (Mueller et al. 1993).

few[34].

We claim RNA editing is a computational mechanism

The point: this process in *T.Brucei*, specifically, the "guide RNAs," is remarkably similar to the action of the "transition matrix" in our Turing machine construction. Processes like RNA editing are thus capable of Turing universal power and may have evolved specifically to cope with the computational demands of gene regulation in eukaryotes. See §3.6 and §3.5.

Now in fact, *T.Brucei*'s particular choice of gRNAs is not Turing universal. This is because the U inserts and deletes always lie on the 5' side of the recognition site. This point seems most clearly made in the Maslov-Simpson 1992 paper. *T.Brucei*'s gRNAs seem in general to act on mRNA sequentially and undirectionally, with each editing step often setting up the recognition template for its successor. Also, one is handicapped by the fact that the poly-A tails are on the wrong side (the 3' side). Full Turing behavior would want to allow bidirectional movement and the capability to add an unbounded amount of extra tape.

Probably this choice of disallowing backward branching by *T.Brucei* was "intentional" because full Turing power would have made it difficult to prevent "software bugs" from getting out of hand and would make things much less predictable.

However, it still remains possible that by combining the RNA editing systems of two different organisms (e.g. a *Physarum* and a trypanosome), that we could in fact get a set of string rewrite rules of full Turing power in a single test tube[35]. Cf. §3.6.

---

[34]There is also some evidence (antisense RNAs) suggesting that edited mRNAs in *T.Brucei* can be *replicated*. It is also known that RNA site specific endonucleases and RNA ligase are around (B. Sollner-Webb, Nature News and Views 356 (1992) 743.). In other words, the RNA editing processes that we know about may not be the only ones in town.

[35]Also, even unidirectional rewrite rules can have impressive computational power if applied to circular RNA – which exists in nature, see Ford & Ares 1994 and Tsagris et.al 1991.

We claim RNA editing exists because either it is a computational mechanism, or it is a relic of one

It has been argued by Landweber and Gilbert 1993 that RNA editing in *T.Brucei* and kin may represent an adaption designed to increase mutation rate and evolvability. Since both mutations in the guide DNA and the cryptogene DNA will cause a mutation, the mutation rate for edited genes, is upped by a factor ≈ 2 (confirmed experimentally by L & G).

We regard this evolvability explanation for RNA editing as illogical. Many biologists have argued (and correctly, as surely L & G would agree) that the reason for various biochemical objects, e.g. DNA repair enzymes, was to *decrease* the mutation rate! If you want to argue that increasing the mutation rate is desirable just in this case, you had better present an evolutionary reason why, for example, *T.Brucei* should particularly want to evolve its cytochrome C oxidase gene more quickly than any other one. L & G gave no such argument. And it is silly to think so. If there is anything an obligate parasite like *T.Brucei* should want to evolve quickly, it is not the internals of its mitochondrion, but rather its external cell membrane – precisely the things that are not coded from mitochondrial DNA. If there is anything *T.Brucei* should want not to mutate, it is cytochrome C oxidase, which (see Stryer) is extremely conserved throughout all known eukaryotes, and in fact is apparently always functional on cytochromes from other species.

So instead, it seems to us that there are two possible explanations for this extensive RNA editing. The first is that it is simply an evolutionary relic. The second is that it serves some important computational function within the cell, the most obvious guess being a control function. Evidence: It is known that life cycle changes in *T.Brucei* are correlated with abundance changes in cryptogene RNAs at different stages of editing: and in the human bloodstream *T.Brucei* turns off mitochrondrial respiration but activates it in insects, although RNA editing is present during both life stages[36] (Stuart 1991).

It is known that gene regulation is a very important and difficult computational problem faced by the cell, cf. §3.5. A mechanism with Turing universal (or nearly), and also potentially highly parallelizable,

---

[36]This mitochondrial switching another unusual feature of T. Brucei that distinguishes it from most eukaryotes, but it may not be the reason for RNA editing, since, e.g. *T.Evansi* has RNA editing but no functional mitochondrion.

computational power could therefore be a useful thing to have right smack in the middle of the information highway.

As far as the first explanation (evolutionary relic) is concerned, it is known that the kinetoplastids are one of the two evolutionarily oldest lines of eukaryotes (according to DNA divergence evidence, Sogin 1989) and RNA editing in them seems to be "very old indeed, perhaps dating to prebiotic times..." (Benne 1994 p.20). So perhaps, this has something to do with the famous hypothesis of early "RNA life" (§3.2).

RNA editing in T.Brucei is a Turing machine and this fact may be exploitable to make a fast in vitro computer

To summarize the last two subsubsections: RNA editing is a natural way to construct a Turing universal computer, and its location in the biochemical information pathway is a natural place for *T.Brucei* to want one.

Now, Seiwert & Stewart 1994 have managed to prepare a *T.Brucei* mitochondrial extract, containing intact editosomes, and used it to do one particular edit step of an synthetic mRNA in vitro. In the event that one could clone editosomes, and also in the event that gRNAs really are a general purpose specification for arbitrary edit operations as is currently thought, we could, by placing editosomes, an energy source (ATP and UTP, probably) gRNAs, a poly-adenylatase and TUTase, and a mRNA "tape" in a test tube, get a second way to make a Turing machine in vitro.

This seems likely to work, although since it is obviously undesirable to work with the organisms that cause African sleeping sickness, it may be best to switch to another trypanosome. This Turing machine, which would really be better describable as a rewrite rule system (cf. §3.6), would have two advantages over our construction. First, it would probably be a lot faster: although we are unaware of any measurement of editing speeds in *T.Brucei*, if one assumes that *T.Brucei* mRNAs have about the same halflife in vivo as mRNAs in other eukaryotes (a few minutes), then editing at 158 sites is performed per mRNA in a few minutes, for a rate of about 1 edit operation per mRNA per second. That is 10^3-10^4 times faster than our (& previous) constructions with present day standard lab techniques, and would allow $\approx 10^5$ rewrite operations per day per mRNA. Second, since the underlying computational model would be a string rewrite system (see §3.6) rather than a Turing machine, additional parallelism would be obtainable. But unfortunately, since RNA is less stable than DNA to backbone hydrolysis (Lindahl 1994) especially in the presence of ions such as Mg^{++}, this computer would dissolve even more quickly than our DNA based construction.

3.2 "RNA life" ideas of the origin of life

The orthodox picture of present day life has DNA, which carries the information and replicates, RNA, which transmits the information from the DNA to the ribosomes, ribosomes, which synthesize proteins according to the commands on messenger RNA, and proteins, which do the work of the cell – whether it be chemical (enzymes), mechanical (actin and myosin; microtubules; cilia) or structural. There are also lipids, which form the cell membrane, but their synthesis and destruction is overseen by proteins. Finally there is ATP (and sugars), which are the currency of energy in the cell. All these components are highly specialized and optimized for their tasks.

In the early days of life, though, things could not have been this sophisticated. One well known speculation (Miller & Orgel 1973, 1974; Joyce 1989) is that in an earlier stage of life there was just RNA. The RNA played both the genetic role currently played by DNA, and the enzymatic roles currently played by proteins, and as for an energetic role, of course ATP is highly related to RNA. Since RNA viruses exist, it is known that RNA can serve to carry genetic information. Recently, pure-RNA enzymes (Kruger 1982, Guerrier-Takada 1983) have been discovered (the tRNAs, ribosomal RNAs, and telomere template RNAs had been known for a long time to be co-enzymatic, Cech 1994) so it is known that RNA, unassisted, can play enzymatic roles. (What about lipids and the cell membrane? Not clear. Perhaps originally no membrane was needed since life was just a single replicating RNA molecule?) RNA can form complicated structures stabilized by many self-hairpin pairings. Nucleotides and their phosphates (e.g. ATP) are known to arise in "primordial soups" (Joyce 1989). Recently, evolutionary selection has been repeatedly demonstrated by an in vitro, selective and error-prone, RNA replication process, leading, after 10 generations starting from a random sequence pool, to a multimillion-fold enhancement in the particular enyzymatic activity being selected for (Ellington & Szostak 1990, Lorch & Sostak 1994, Sassanfar & Szostak 1993) including the de novo synthesis of an all-RNA RNA ligase (Bartel & Szostak 1993).

Pace 1991 has pointed out that all-RNA life would have been very unstable under the severe hydrolysing conditions that probably existed over most of the primordial earth, and therefore perhaps "RNA world" was preceded by some earlier and even simpler form of life; a stabler artificial polymer called "PNA" (peptide nucleic acid, Nielsen et al. 1991) may fit into that picture.

Anyway, it has been shown that RNA molecules, acting alone, are capable of catalysing RNA cleavage and joining[37] (and self splicing, self cleaving, and self circularizing; see Cech 1987, Kruger 1982, Uhlenbeck 1987). An entirely enzyme free system has been shown to be capable of "high" fidelity RNA polymerization guided by RNA templates 6-10bp long (Inoue and Orgel 1983; admittedly no such system has been demonstrated with racemic nucleotide mixtures).

The idea that one can make an RNA Turing machine should be important for "RNA life" ideas of the origin of life

The connection between all this and RNA editing and our Turing machine is just this: we will demonstrate in §3.6 that it is easy for systems capable of RNA editing like operations, to get Turing universal power. With all this evidence, it seems quite possible for an all-RNA system to evolve with Turing universal power. Thus, computer science may in this case be providing insight into how life could have arisen.

3.3   Introns, Transposons, and junk

Transposons are subsegments of your DNA which can transplant themselves from one location within your genome to another. *Replicant transposons* (which we will call "replicons") can do more than just move – they can in fact make copies of themselves which implant themselves elsewhere in your genome. *Introns* are substrings of your genes, 10^1-10^4bp long, which are transcribed into mRNA but then excised by *spliceosomes* before that mRNA is used to direct the manufacture of proteins. (The non-excised portions are *exons*.) 97% of the human genome is *junk DNA* which has no known function – and since much of it consists of short repeating sequences, it cannot be encoding very much information.

---

[37]Incidentally, DNA in single stranded form does not exhibit catalytic activity – Cech: "the importance of being ribose."

All of these things seem to be at best a large waste of resources and at worst fatal. Fatal because it is known that transposon hits can deactivate vital genes and thereby kill the cell – and more than half of the observed mutations in *Drosophila* are due to transposon hits.

Indeed, replicons seem to be taking over the genomes of higher life forms. 10% of the human genome is currently made of copies of the two most popular kinds of replicons, called "Alu" and "L1." There are 10^6 copies of Alu, and each one apparently serves no function other than directing the synthesis of proteins important for its own copying!

Meanwhile in bacteria such as *E. Coli* there is almost no junk, few if any introns, and the only known replicons are phages (such as μ-phage) which tend to be fatal and are not present in healthy bacteria. Even *E. Coli* has ≈ 10 transposons, however. Some of these transport themselves to apparently random sites, some to hotspots, and some only to sites with special recognition sequences such as NGCTNAGCN. The transposons in *E. Coli* seem to move at rates of 10^{-4}-10^{-3} migrations per transposon per cell per generation.

Transposons can be "autonomous," which means that they encode all the proteins necessary to carry out their migration and/or replication or "nonautonomous," which means they have to parasitize proteins from other parts of the geneome, possibly including other transposons. All replicons known so far use a reverse transcriptase, similarly to μ-phage. Replicons may be viruses which have lost their coat protein genes and consequently the ability to exist outside the cell (although the other hypothesis is that viruses evolved from replicons!) and nonreplicons may be replicons which have lost the ability to multiply exponentially, but still persist through evolutionary time because they confer some advantage to the host.

Why do replicons, junk, etc. exist? The old explanations and their problems

What evolutionary advantages, if any, do transposons and junk confer? Why are they there? Transposons can have uses. Genes located between two transposons (or within a transposon) can get moved around in the genome, and also copied, and similarly exons can get shuffled by transposon action. On evolutionary time scales, the combination of transposon and introns can provide a mechanism for generating useful mutations. A species without transposons and only evolving by point mutations might take virtually forever, say, to evolve three different rhodopsin genes in order to acquire 3-color vision. It would be out-evolved by a creature whose replicons happened to make three copies of its rhodopsin gene, and then perhaps some of the exons got shuffled by nonreplicant transposon action... and the three rhodopsins then, finally, got tuned by point mutations to be sensitive to different colors. The point is that reshuffling and copying gene building blocks which are known to be useful, is much more likely to evolve a useful new gene than random point mutations alone. Indeed, creatures that reproduce sexually go through a stage called "crossing over" in which, conceptually, substrings of the maternal genome are swapped with corresponding substrings of the paternal genome, with the boundaries of the swapped regions apparently being chosen at random. The crossover process undoubtably arose because it causes variability in the genome which has a comparatively large chance of being useful and a comparatively small chance of being harmful. Perhaps the evolvability advantages of creatures whose genomes were equipped with transposons, outweigh the dangers.

On much faster time scales, of a single lifetime, transposons, or mechanisms similar to them, also can have uses. Transposons are suspected to be part of the developmental and regulatory process in the maize plant. (Obviously, by moving promotor and repressor subgenes around, regulatory and developmental effects are achievable. Indeed, the nematode *Ascaris* reshuffles its whole genome during development.) As discussed in §3.5, reshuffling of DNA segments is an essential part of antibody synthesis in vertebrates. The amplification of genes made possible by mechanisms like replicons causes certain genes in *Drosophila* salivary glands (presumably genes related to making saliva) to be present in over 1000 copies. These genes are present only in much smaller numbers in normal *Drosophila* cells. Similarly, *Drosophila* ovary cells have genes which are greatly amplified.

Do not be convinced that transposons and introns arose because of all the advantages they confer. Amplified genes are also known to be present in mammalian cancer cells, which suggests that this mechanism is dangerous – cancer is a leading cause of death in humans. Human leukocytes have a higher risk of cancer than other human cells, which is probably related to the fact that they shuffle their genomes (cf. §3.5). *Drosophila* males are known to have lifespans that are shorter, the more P replicons they have (Woodruff 1992, see also Driver & McKechnie 1992), presumably due to an increased load of somatic mutations. Meanwhile it is known that silk moths, some of whose cells produce huge quantities of silk protein, do *not* amplify the silk gene, which indeed is present in only one copy. In fact most eukaryotic genes are present in only one copy. That suggests that gene amplification, while perhaps useful, is not necessary. There is also some evidence that exons are not especially useful gene building blocks and that the intron borders occur rather randomly, but all such evidence so far is hard to interpret and cannot be said to be conclusive.

Finally, the biggest problem of all: replicons are very bad news. It has been observed that the most successful replicons (e.g. L1) have negative feedback mechanisms that prevent them from multiplying too rapidly. (Still, their growth is assuredly exponential over evolutionary time scales.) This is because any transposon which multiplied too rapidly would take over the genome and kill off the species. But, suppose that the replicon multiplies slowly enough that it can contaminate the genome of every member of that species, before it grows large enough to have any injurious effect. (Alu has managed that.) In that case, that species seems automatically to be an evolutionary dead end. And it furthermore seems to us that such replicons will always be there, since evolutionary processes favor the replicons with the highest growth factors that are possible without death. So, replicons ought to make higher life forms impossible.

A "computer science explanation" for replicons and junk

Here is our interpretation of all this. Higher life forms had to have an immune system (cf. §3.5), so they had to have gene reshuffling or some other highly computation-like ability in order to outcompute their simpler parasites. So Turing universality was inevitable – not only for this reason, but also because of the huge evolvability and developmental/regulatory advantages that it allowed. But once the necessary mechanisms for Turing universality became widespread, life became very dangerous. As is well known, it is very easy to change a computer program to make it do something unpredictable, nonlocal, complicated, and possibly destructive. In particular, replicant computer programs are possible and, as is also well known, can be very dangerous – and that is exactly what replicons are. The same sort of gene shuffling mechanisms used by vertebrates to fight infections are known to be used by eukaryotic parasites to synthesize coat glycoproteins which keep changing. There is thus a computational arms race between hosts and parasites and with the advent of Turing universality it in fact becomes undecidable whether, e.g. something inside your genome is harmful... So for this reason, there was incentive for cells to try *not* be be Turing universal, or at any rate to try to keep their computational capabilities under as tight control as possible.

We have argued that replicons, if left to themselves, ought to make higher life forms impossible. But higher life exists. It seems to us that the only possible solution to this paradox is to infer – or predict – that replicon "police" exist. Suppose some replicon were to start becoming a heavy burden, occupying, say, 99.9% of the genome. Obviously, if some "killer" gene were to evolve which produced enzymes which recognized that replicon, excised it from the genome, and destroyed it, then members of the species which had the killer gene would have an advantage and the replicon would be wiped out. On the other hand, once the replicon was gone or greatly reduced, this killer gene would be a very dangerous thing to have, since it produces DNA destroying enzymes. It would quickly vanish[38]. Because the most prolific transposon would tend, as it grew exponentially, to destroy its opposing less prolific transposons (but not destroy essential genes), over evolutionary time, by the time it became necessary to call in the police, only one replicon would desperately need to be killed, and the cure therefore would work.

Unfortunately, the replicons can fight back. One way is to camouflage themselves as essential genes. The Alu replicon strongly resembles the essential 7SL gene, which is usually interpreted as evidence that it evolved from that gene, but could also be interpreted as the self defense measure that explains why Alu is still around.

We further speculate that junk DNA may be in large part composed

---

[38]That is perhaps why no replicon-killer gene has been seen so far and this is all speculation. But it is still provable. If it were shown that the Alu replicon were not present in monkeys, that would prove it appeared after the divergence of humans from monkeys – and grew to 5% of the genome in that comparatively short time. The only way such growth rates could be possible and higher life forms could still exist, would seem to be killers. Actually Alu variants exist in many mammals, although not in frogs and flies even though they have similar 7SL genes, and appear to have arisen about 65 Myrs ago (Deininger & Daniels 1986). This still seems sufficient to make our case, since (1) the Precambrian metazooan "bio-explosion" occurred $\approx$ 530 Myrs ago and (2) Alu's proliferation from 1 to 10^6 copies per genome in 65 Myr implies a doubling time of 2 Myr, which seems to leave the human genome with at most 9 more Myr to go before Alu takes over 100%. It is also known that the P replicon in *Drosophila* has appeared, and multiplied by a factor 2-50, in only 50 years.

of the corpses of former replicons, which would explain (1) how it got there, (2) the fact that a lot of junk is highly repetitive (Britten and Kohne 1968)[39] and (3) our theory is also supported by the fact that closely related species often have vastly differing amounts of DNA (factors of 5-10).

Finally, even with replicon killers, higher life forms still face the problem that their genomes may gradually become overwhelmed by junk. Oddly, it may be that nonreplicons may provide a way to keep down the junk, since such a transposon could, by failing to reinsert itself, serve to delete a piece of DNA. If the deleted segment were junk, this deletion would be harmless, indeed slightly beneficial.

3.4   The secret lives of hypotrichous ciliates

Hypotrichous ciliates (single celled eukaryotes such as *Euplotes Crassus*, see Prescott 1992 and Tausta et al. 1991) have two micronuclei (diam$\approx 4\mu$m) and two macronuclei (diam$\approx 25\mu$m). Each micronucleus contains what at first appears to be a fairly typical looking eukaryotic genome.

However, following mating (that is, exchange of a pair of haploid micronuclei which fuse to make normal diploid micronuclei; this is comparatively rare, usually reproduction is asexual), all nuclei are destroyed except for the new micronucleus, which is copied. The new micronucleus then gives rise to a new macronucleus. During the construction of the new macronucleus, remarkable events happen. The DNA from the micronucleus is chopped into a large (5×10^4) number of short (smaller than a gene) segments; some of these segments are removed, some are destroyed, and some are ligated – often in a different order. For example, the 1.5Kbp actin I gene is cut out of the genome and chopped into 17 pieces. Three of these pieces (*"bcd"*) join to form a 175bp circle, one (*"a"*) forms its own 23bp circle, and the remaining 13 pieces are permuted according to

$$\ldots 3a4b6c5d7e9f2g1h8 \ldots sf987654321ghe,$$

where s is a 15bp spacer which is apparently synthesized de novo, to

---

[39]It has also been speculated that such short repetitive sequences as 5'-ATAAACT-3', which form 25% of *Drosophila* DNA, may serve some structural function, but this function is probably not very important since even closely related species often have totally different repeats.

get a correct (transcribable) actin I. (The original pre-permuted actin I gene wouldn't have worked for directing the synthesis of actin.)

After such operations are complete, and huge amounts of non-gene DNA (i.e. $\approx 95\%$ of the genome) are destroyed, one is left with a large number of DNA segments (200-15000bp), each of which contains a single transcription unit.

These are then equipped with standardized telomeres which are also synthesized de novo, and then amplified by ≈ 1000 fold replication. Finally, the finished macronucleus divides into two macronuclei, each of which contains about 2×10^4 gene-sized species of DNA, each species coming in ≈ 1000 copies, and the cell continues its merry way, with two macro and two micronuclei. The macronuclei are used for all transcriptional purposes and the micronuclei for apparently no purpose, except for storing the genetic information for eventual use during mating and the production of new macronuclei.

How and why[40] is all this done? All we can say is, this is quite amazing.

3.5   The cell as a computer

Cells are known to face and solve some difficult computational problems.

Antibody synthesis

There are $\approx 10^{12}$ immune system cells in your body, which is at least an order of magnitude more than the number of your neurons. Arguably your immune system cells are doing a lot more computation than your brain.

What does your immune system do? It detects foreign substances, ("antigens") and then designs, de novo, an antibody (protein) molecule that binds to that antigen. Antigens include "any inorganic or organic molecules you care to mention... when attached to a [suitable]

---

[40]Our conjectural explanation for "why" is that this is a "preprocessing" step. In computing it is well known that time spent beforehand doing a precomputation (e.g. sorting a phone book) will often more than pay for itself later. Similarly, by preprocessing its DNA to remove introns and junk, *E. Crassus* is able to avoid the later metabolic cost of manufacturing all the intron mRNA which is wasted by other eukaryotes. By retaining introns in its micronuclear genome it still gets whatever evolvability advantages they confer.

carrier molecule," according to Jerne 1984, and that includes molecules just synthesized by chemists which have never existed before. Once the design process is completed, the antibody is synthesized in comparatively large quantities, circulates throughout your body, and binds to antigens. Killer "T" cells then rove throughout your body, find things which have antibodies attached to them, and kill them, e.g. with digestive enzymes. Similar systems are in all vertebrates.

Many details of the design process are now understood[41], and here is the rough picture.

Antibodies consist of a standardized Y shaped portion and variable portions near the "hands" of the "Y." The variable portion is coded for by a DNA region containing some of the 200-300 antibody building block subsequences. "Resting B-cells" use recombinases to continually shuffle their antibody building block subsequences into and out of the coding region. The recombinases recognize a specific heptamer and nonamer upstream and downstream of each building block. With 200 blocks, of course $\approx$ 200! permutations would be possible, but even more variation is possible since the building blocks can also get shifted within prescribed ranges and immune cells also exhibit point mutations (in this DNA region) at a much higher than usual rate ("hypermutation"). Thus a region of the DNA of B-cells is continually changing. If and when a moment occurs when this DNA happens to code for antibody which binds to the antigen, the B-cell is stimulated to cease exploring DNA-space and instead to reproduce. Its descendants serve as factories each outputting 2000 antibody molecules per second. But some fraction of these descendants continue to explore DNA-space in order to fine-tune the antibody's antigen affinity; thus the antibody design and synthesis process is an example of *Lamarckian* evolution going on inside your body in an effort to solve a combinatorial problem by an exponentially large heuristic search.

Most simple parasitic organisms cannot compete with the huge

---

[41]The related problem of how the immune system learns what substances are "friend" and which are "foe" is not understood. Antibodies themselves can serve as antigens, leading to the production of anti-antibodies. At any given time there are $\approx 10^7$ different kinds of antibodies, totalling $\approx$ 100grams, circulating in your body, as compared to the $\approx 10^5$ proteins coded for in the genome we were born with, and (Jerne 1984): "...in its dynamic state our immune system is mainly self-centered, generating anti-idiotypic antibodies to its own antibodies, which constitute the overwhelming majority of antigens present in the body. The system somehow maintains a precarious equilibrium with the other normal selfconstituents of our body, while reacting vigorously to invasions..."

parallel computational capability implied by our 10^{12} immune cells, which is why we can recover from many diseases.

Lamarckian evolution will also rear its head in §3.5.

Gene regulation

The prokaryotes (cells without nuclei) seem to have an upper limit of about 6000 genes, and this may in part be due to the problems inherent in controlling that many genes with the limited mechanisms (repressor and promotor proteins) available to them. Eukaryotes have many more genes than this. Humans are estimated to have 10^5 genes, and there are at least 5 regulatory elements per gene. (Robert Tijan, quoted in Science 263 (Feb 1994) p608.)

Certainly the usual promotor and repressor (proteins binding to DNA near the gene they regulate) mechanisms are used extensively by eukaryotes as well as prokaryotes. The *Drosophila* Eve gene has a 2×10^4bp prologue, used for regulatory purposes, which binds > 20 regulatory proteins, and this is small potatoes compared to some mammalian genes. But what do humans and other eukaryotes have that prokaryotes do not, that allows their cells to handle the huge excess regulatory burden?

One answer may well be: introns. Prokaryotes cannot use intron splicing since they have no nucleus to physically separate incompletely edited RNA from ribosomes. It is known that an intron in the roundworm *C.elegans* codes for an RNA that binds to and deactivates the mRNA for another gene. It is also known that certain aminoglycoside antibiotics deactivate certain RNA intron splicing processes (von Ahsen et al. 1992), which suggests that these are important processes. The fact that intron excision is done by a multiprotein spliceosome complex also suggests its importance. Nevertheless, at the present time, most introns have no known function and ignorance is widespread.

In principle, RNAs could bind to mRNA introns, and excised introns could bind to other mRNAs, and some of these bindings could either deactivate or activate spliceosome or RNAase actions. This could set up a complex network of possible interactions which could serve a computational and regulatory function operating on fairly short time scales. (The half life of mRNA in rat liver cells is known to be a few minutes.) This is the usual sort of speculation about the possible function of mRNA splicing in eukaryotes. What is the computational power[42] of such a network?

Assembly of complex structures; development of multicellular organisms

How can it be that all of our cells have the same DNA sequence[43] but we have many different kinds of cells? How do our muscle cells, retinal cells, nerve cells, bone cells, fat cells, kidney cells, and so forth[44] know which to be? How do they all manage to assemble themselves into a coordinated structure? Neurons are single cells whose axons can be meters long, and these axons somehow thread themselves between all your other cells to go to the right places. All this is the problem of multicellular development and differentiation, one of the "great unsolved problems" in biology.

A perhaps related question is: how do complicated structures *inside* cells get assembled? It is known that if the component parts of some virus particles or ribosomes are placed in buffer, these complex assemblies will magically self-assemble. Many proteins magically self-fold into the correct shape. But more complicated cellular structures will not self assemble from their components. For example, entire cells will not re-assemble, after they have been destroyed... and so far as is currently known, nor will complex subcellular structures such as cilia, mitochondria, and centrioles.

As far as the assembly of complex subcellular structures is concerned, simply providing the components in the right temporal order, rather than all at once, may go a long way. This is something that finite state machines, or the more powerful models of computation we describe in §3.6, are easily able to do. Similarly to the way that construction crews assemble large buildings, temporary "scaffolding" may be required which is later destroyed. Thus it is known that some bacteriophage heads are assembled with the aid of a second protein which does not appear in the final phage heads, which therefore won't reassemble if they are disassembled (Mathews et al. 1983). It may also be that the transportation system in the cell has to be more sophisticated than is currently supposed. Similarly to the way that

---

[42] AND, OR, and NOT functionalities would seem thinkable which would imply "Boolean completeness."

[43] Or at least, that is the orthodox picture; of course, our immune system and the phenomenon of gene amplification are counterexamples to this picture.

[44] Alberts et al. 1994 give a catalogue of 210 major categories of human cells.

ribosomes are general purpose protein constructors which work to order to mRNA blueprints, the microtubule vesicle-transportation system of eukaryotic cells may also work to order, according to some sort of "train tickets." It is a commonplace that membranes in cells have many sophisticated kinds of portals, and it is known (Lombardi, Soldati, Riederer 1993) that vesicles "dock" with the right lipid membranes with the aid of recognition molecules called "Rab proteins."

Be that as it may, the entire problem of multicellular development is easy to explain, assuming we live in a certain oversimplified mathematical model universe – a "cellular automaton." See §3.6.

Protection and Lamarckian evolution

The orthodox ("Darwinian") view of evolution is that mutations are entirely random, and the ones which happen to confer increased fitness, are the ones which stick around. "Lamarckian" evolution, is the hypothesis that organisms somehow sense what changes need to be made, make them – more or less "intentionally" – and then pass them on to their offspring. It is clear that Lamarckian evolution, were it feasible, would confer a great Darwinian selective advantage... so that what one should expect is some compromise between Darwin and Lamarck. While it may not be possible to choose intentionally which mutation you want, in many circumstances it may at least be possible to bias the Darwinian dice to make mutations which are more likely to be favorable, more frequent, and mutations which are more likely to be deadly, less frequent.

Meanwhile in computer engineering, it has long been agreed that, in order to build robust computing systems, varying degrees of "protection" and "damage limitation" are essential. The "operating system" and its work areas are generally highly protected – by hardware – against modification by user code. Indeed, some machines have varying levels of protection within the operating system. Also, certain useful things the computer can do, e.g. disk I/O, user programs are *not allowed* to do – unless they work through calls to operating system subroutines which decide whether they will deign to permit it. All this protection in fact *reduces* the accessible computational power of the machine and the creative freedom of the programmer, but it is nevertheless regarded as essential for robustness in the real world of bug-filled programs.

By analogy, we argue that there ought to be varying degrees of protection against various kinds of mutations in different parts of your genome. The most essential and evolutionarily oldest parts ought to be the most protected, and the parts which seem most likely to have something to do with your current problems (e.g. highly expressed?) should perhaps be more mutable. In addition to this sort of varying spatial protection, one would want varying temporal protection – the best strategy is to choose to mutate more when you are under selection pressure. Of course, when one is under such pressure, one may be *forced* to mutate more – a starving lifeform may not have the resources to correct genome damage and conduct high accuracy replication – but it would be better to try to mutate intentionally before this point is reached, because that way you will be more able to control both the rate and the type of mutations you will incur.

Suffice it to say, there is evidence in support of all of this. We have already argued (cf. §3.3) that transposons and sexual evolution (and also genome "crossover") may have arisen in order to make more favorable mutations more likely. The fact that organisms under selection pressure mutate more, and in particular have more transposon mutations ("transposition bursts"), which are in some sense "intentional," is well known. (And 80% of *Drosophila* mutations are due to transposon hits.) Genes, such as the famous p53 gene, exist which can exert control over the transposon rate – when p53 is turned off, which often happens in cancerous cells, this rate can increase by a factor of 10^5. It is known (Kricker, Drake, Radman 1992) that methylated dimers can serve to protect DNA regions, in particular important mammalian genes, against transposon hits.

There is some evidence, though not yet as convincing as the above, that more important genes mutate less often during replications. Especially, there is evidence that *E.Coli* with defective lactose processing genes, will mutate more often, in different ways, and more than usual in the Lac gene region when exposed to lactose selection pressure. When stressed in the absence of lactose, this bias vanishes. Some recent experiments by Cairns and by Foster have been very convincing in this line (Foster & Trimarchi 1994; Cairns, Overbaugh, Miller 1988; Cairns & Foster 1991, 1992).

3.6   Models of computation in vivo

Orthodox bacterial model

A computational model we dreamed up to encapsulate and simplify the internal operations of bacteria (or at any rate, the orthodox picture thereof), say, is as follows.

Bacterial model of computation.

1. You have an infinitely long "tape" which is read-only. Each tape square contains one of 4 characters. In addition, characters may be of two colors, red or blue. The tape contains the "program" and "data" and its initial state could be encoding an infinite amount of information, but let us say that initially only a finite number of readable substrings of the tape are red.

2. A computational step is: you find a all-red substring S of the tape which is long enough to contain a special "start" and "end" subsequence. This substring specifies another substring S_2 and one of two actions to take:

A Find all substrings in the tape which are the same as S_2 and which are currently all-red, and color them all-blue.

B or: Find all substrings in the tape which are the same as S_2 and which are currently all-blue, and color them all-red. (Substrings which are partly red and partly blue are unaffected.)

3. The rules are: if a computational step is possible, pick one of the possibilities at random and do it. Keep going forever (or until no step is possible; noninterfering computational steps could be run in parallel).

Obviously, the "tape" represents the bacterium's DNA, the 4 symbol alphabet corresponds to the 4 nucleotides, and the two colors blue and red correspond to "bound by a repressor protein" and "not." The fact that the tape is read-only is the orthodox picture of bacteria.

The question is, what is the computational power of this strange model. If the tape is finite, the situation is boring from the computer science point of view: the situation is just a finite state machine.

If the tape is infinite, Hava Siegelman showed us rather simply that the bacterial model of computation would have super-Turing power. The proof is simple. Prewrite, in order, an uncomputable sequence on the tape, in the format $1, f(1); 2, f(2); 3, f(3); ...; n, f(n); ...$ where $f()$ is some uncomputable function. The ; 's contain start and end sequences and the integers n are colored blue initially. Then, to read out $f(n)$ in two steps, paint n red in step 1 and read $f(n)$ in step 2.

Of course this super-Turing power is not really accessible biologically, but this little example does start the mental wheels turning.

With a rather more simply preformatted tape, namely just a periodic sequence, we still claim that fully Turing power is possible, *provided* that the model is extended so that binding proteins will only bind to (or be removed from) their DNA recognition sites *if* said site is sufficiently close to a previously bound protein which is a promotor of such activity. ("Eucaryotic gene regulatory proteins often assemble into small complexes on DNA" – Alberts et al. 1994.) We claim that a finite number of such proteins and relationships among them will suffice to emulate a universal Turing machine and with only constant factor slowdown.

The proof is omitted, but straightforward. For example one can construct an emulation of a Post "tag system" as in Minsky §14.6.

A computational model based on RNA editing

Suppose we have mRNA, which we model again as a tape with 4 symbols, and we have a finite set of rewrite rules which specify certain context specific single nucleotide conversions, insertions, or deletions, or which can splice out an entire intron (substring) provided it has specific recognized start and end sequences, in one step.

Any such system of rewrite rules may have ambiguities, i.e. several rules may be applicable at any time, in which case we break ties temporally at random.

Again we claim this model of computation has full universal Turing power. (Of course, this model is extremely similar to Chomsky grammars and Post rewrite systems.)

Although all of these editing capabilities are known to exist biologically, they are not presently known to exist all at the same time in the same organism. To make matters realistic, then, we proclaim that in fact,

> Only a finite collection of rewrite rules which insert or
> delete single U's in specified contexts of an mRNA sequence
> will suffice to get Turing universality;

no nucleotide conversion rewrite rules, or insertion or deletion of non-U's, are needed. This is compatible with current knowledge of *T.Brucei*. We do need an infinite mRNA "tape" however, prewritten with a certain simple periodic pattern (poly-A will suffice) except in some finite portion, or equivalently we need the capability to append an unbounded number of A's, say, to one end of our mRNA. (Yes, only A's and U's are needed.) Eukaryotes are known (e.g. Baker 1993) to have mechanisms for appending long poly-A tails to the 3' end of mRNA.

The proof of this can be accomplished by embedding a 1-dimensional Turing universal cellular automaton with a two character alphabet and a neighborhood notion 18 bits wide in the *T.Brucei* model, where the tape is a periodic sequence in which, in each period, a U can either be present or absent in specified locations. Alternatively one can construct a universal Turing machine directly with certain special head states and tape square states.

A cellular automaton modeling multicellular development

A "cellular automaton" (CA) is a model of computation in which the world is an infinite square (in 2D, or cubical, in 3D, or at any rate periodic) grid, and each square contains a single character from a finite alphabet. All cells are updated synchronously to become, at time $t+1$, a new state which is a function (the "transition rule") of its previous state at time t, and that of its immediate neighbors.

The "game of life" is a 2D cellular automaton invented by J.H. Conway, described by these incredibly simple rules: cells have two states, "alive" or "dead," and live cells stay alive at time $t + 1$ if and only if they have 2 or 3 live neighbors at time t, and a dead cell turns alive if it has exactly 3 live neighbors. In Conway's book (Berlekamp et al. 1982) the game of life is shown to be capable of Turing universal power[45].

Biologically, we may state this as follows: **Cellular automaton model:** at any time, depending on its internal state, that of its immediate neighbors, and an internal random number generator, a cell may choose, according to a finite set of allowed transition rules, either to divide in a specified direction, or to change state, or to die (vanish).

---

[45]Admittedly their game of life Turing machine emulation has an exponential slowdown ratio, but almost certainly the slowdown need only be polynomial and this was only an artifact of Conway and co.'s desire to keep the proof simple.

We claim that this model, entirely unaided, is easily capable of duplicating the complicated sorts of differentiated behaviors found in real multicellular lifeforms. Unlike Conway's game of life, though, the cellular automata we are concerned with here have potentially very large state and transition rule sets, since the only constraint – that their specification be small enough to fit into a eukaryotic genome – is rather weak.

We claim that, if the transition rule is sufficiently complex (≈ 40 bits of state information per cell should suffice – a number of bits of course dwarfed by the size of huge eukaryote DNA's), then it is easy to write a program (i.e. state transition rules) so that a single cell, initially alone in the universe, will develop into essentially any complex structure whatever.

Also easily, this program can be made to include self correction mechanisms to make it fairly robust to unintended random cell deaths and mutations.

Testable predictions would include: the existence of messenger chemicals and channels which would allow cells to know about the state of their immediate neighbors (also one could consider extending the model to allow cells to broadcast chemical messengers which would diffuse throughout large regions, but this is not necessary); the existence of crude "clocks" so that cells could keep track of time; and the prediction that a cell in the same state and with the same neighbors will always do the same thing.

Also, the entire transition matrix need not be specified, instructions only need to be specified for those neighborhood structures which are likely to arise in practice. This is perhaps the explanation for "teratomas," which arise from grafting developing fetal tissue into unnatural tissue locales, the result being a "bizarre growth... a disorganized mass of cells containing many varieties of differentiated tissue – skin, bone, glandular epithelium, and so on – mixed with undifferentiated stem cells that continue to divide..." (Alberts et al. 1994).

Are the true mechanisms of multicellular development actually of this sort? Evidence: It is known that some simple multicellular organisms do develop according to rules which entirely fit inside the cellular automaton framework. For example, Mitchison and Wilcox 1972 investigated *Anabaena Catenula*, a simple organism consisting of 1-dimensional filaments of cells of 4 visually distinguishable types, which we shall call $\{A, B, C, D\}$. The transition rules $A \rightarrow BC$,

$C \to BC$, $B \to DA$, $D \to DA$ seem to hold whenever one cell divides into two. "We have tested this rule in more than 600 individual cell divisions and have found that it holds without exception," they wrote.

The entire life cycle of the hermaphoditic transparent roundworm *Caenorhabditis Elegans* has been studied to the point where the entire ancestral tree of every cell's life (each of the 959 somatic cells has a name) is known: every cell divides into cells of a specified type according to a unique schedule. But the situation must be more complicated than this in higher organisms – certainly each of the $\approx 10^{13}$ human cells cannot have its own name encoded by the original zygotic DNA, since that DNA is only 3×10^9bp.

It is known that tissue cells in both insects and mammals have some idea of their position within the body, and cells whose state corresponds to one such location can still behave consistently with their former state if they are artificially transpanted to a different location. Alberts et al. 1994: "One of the most remarkable revelations of modern genetics has been that almost all animals seem to use the same highly conserved... machinery to record positional values along the head-to-tail axis of the body... discontinuities of positional value provoke local cell proliferation, and the newly formed cells take on intermediate positional values so as to restore continuity... the molecular mechanisms that underlie this crucial form of growth control are unknown."

A *morphogen* is a substance whose concentration is read by cells to discover their positions relative to "landmark," or "beacon" cells. Alberts et al. 1994: "Morphogens are thought to be a common way of providing cells with positional information or controlling their pattern of differentiation, although there are still only a few cases where a morphogen has been identified chemically." These are to be distinguished from "hormones," which are transmitted bodywide and whose concentration gradient, if any, is not thought to be of interest.

We disparage the morphogen hypothesis, despite the fact it was proposed by our hero Alan Turing. Certainly morphogens are important for symmetry breaking in early fetal development, and perhaps elsewhere (for example, perhaps the growth of capillaries toward oxygen-poor tissue is stimulated by a morphogen secreted by such tissue, or perhaps oxygen itself is the morphogen). But, morphogens are inherently inaccurate and error prone, because they are analog rather than digital. They are also incompatible with the sharp digital boundaries that appear everywhere in organisms (e.g.

the surface of a bone). Threshhold mechanisms have been proposed by
morphogen fans to get around that problem, but it isn't that simple...
such problems do not arise in the cellular automaton picture.

The positional-address genes, or at least, the few that are currently
known, are digital.

Consider neuronal axon growth – axons grow along a precisely
defined path, sending out and retracting filopodia to explore the regions
ahead before deciding on their growth path. (Later neurons often
simply follow parallel to a trailblazing axon, though, resulting in a
neuronal bundle.) Also it is worthy of note that before sending out
their axons, neurons first migrate to a precisely defined location.

In conclusion, one may conjecture that the mechanisms of
multicellular development resemble, computationally, the "cellular
automaton" picture and that the "morphogen gradient" picture is
inadequate. The above facts may not prove it, but they do make this
plausible.

It also seems very plausible that the algorithms encoded by cells's
DNA must have some sort of "higher level language" allowing entire
subroutine calls such as "build an entire arm" to be encoded in a small
space. The very existence of people who have functioning extra fingers,
say, seems to prove that. So, it seems to us, looking at the problem
of multicellular development from the computer science point of view,
that experimentalists's attention should be devoted to figuring out the
the transition matrix, the chemical messengers, the clock[46], and the
higher level language[47]. It is easy to see how, in principle, to deduce
the transition matrix – simply take cells with various known states, put
them in various neighborhoods, and see what they do. In the case of

---

[46] It is also known that cells can change "state" (in our cellular automaton sense)
as a function of time – they *do* have crude clocks.

[47] From the standpoint of biology, it is known that all of these capabilities exist,
although they are not well understood at the moment, and it is not known if they are
all actually used in the manner we have suggested. Thus "subroutine invocations"
are sometimes known to biologists as "homeobox genes" and other times are called
"genetic switches." Incidentally, we are using "higher level language" here in the
computer science sense (e.g. MODULA, PROLOG) and we wish to disown any
connection to Mantegna et al. 1994. There it was observed that introns in DNA
have Zipf power law histograms, and various deep-sounding linguistic conclusions
were drawn. Unfortunately Mantegna and his 6 coauthors were unaware that words
of either fixed or variable lengths whose letters are drawn entirely at random from a
fixed alphabet with fixed unequal letter probabilities, also exhibit Zipfian statistics.

the other three things, it is progressively less obvious what to do.

Open problem #4: *Investigate multicellular development from this computer science point of view.*

4 Conclusion

Standard DNA lab techniques may be used to construct molecular Turing machines. Turing machines are simple but universal computers. This is a "nondeterministic" Turing machine with all the enormous parallelism that that implies – up to the point that we run out of DNA molecules. The known in vitro constructions will not be practical in competition with conventional computers, but the reasons for this trace to technological limitations that do not apply *in vivo*.

The view of "biology as a computer programming programming problem" may allow insight into biology and may also suggest new directions in which to extend computer science. For example, why is the biologically bizarre process of RNA editing there? From the computer science point of view we see that RNA editing is a very natural way to get Turing universal computational power, located in a very natural place in the biochemical information pathways. We have also given a natural sounding hypothesis about why junk DNA is there. As an example in the other direction, biology suggests a number of computational models whose mathematical properties are presently little known. Although computer scientists often in the past had been satisfied with proving some model (with an infinite tape) is Turing universal with some (unknown) polynomial slowdown ratio, or as powerful as a stack machine, and so on, it seems that biology may now for the first time be providing incentive to make finer distinctions than these.

5 Acknowledgment

Allan Schweitzer was a major contributor to this paper but wished not to be recorded as an "author."

Bibliography

[1] L. Adleman, Molecular Computation of Solutions to Combinatorial Problems, Science 266 (Nov. 11) 1994, 1021–1024.

[2] von Ahsen, U., Davies, J., Schroeder, R., Non-competitive inhibition of group I intron RNA self-splicing by aminoglycoside antibiotics, Journal of Molecular Biology 226,4 (Aug 1992) 935–941.

[3] Alberts, B., Bray, D., Lewis, J., Raff, M., Roberts, K., Watson, J., Molecular biology of the cell (3rd ed.) Garland 1994.

[4] Andre, P. and Dubois, O., Using the expected number of solutions to optimize the resolution of a SAT instance, Compte Rendu de l'Académie des Sciences de Paris, tome 315, série I (1992) 217–220.

[5] Baker, E.J., 367–415 in Control of mRNA stability (eds. J.G. Belasco and G. Brawerman) Academic 1993.

[6] Bartel, D.P., and Szostak, J.W., Isolation of new ribozymes from a large pool of random sequences, Science 261 (1993) 1411–1417.

[7] Benne, R., et al., Major transcript of the frameshifted coxII gene from trypanosome mitochondria contains four nucleotides that are not encoded in the DNA, Cell 46 (1986) 819–826.

[8] Benne, Rob, RNA editing in Trypanosomes, Molecular Biology Reports 16 (1992) 217-227; Eur. J. Biochem. 221 (1994) 9–23.

[9] Berlekamp, Elwyn R., Conway, John H., Guy, Richard K., Winning ways, for your mathematical plays, Academic Press, 1982.

[10] Blum, B., Bakalara, N., Simpson, L., A model for RNA editing..., Cell 60 (Jan 1990) 189–198.

[11] Bonner, T.I., et al., Reduction in the rate of DNA reassociation by sequence divergence, J. Molecular Biology 81 (1973) 123–135.

[12] Brady, A.H., The determination of Rado's noncomputable function Sigma(k) for four-state Turing machines, Math. Comput. 40, 62 (1983) 647–665.

[13] Brail, L., Fan, E., Levin, D.B., Logan, D.M., Improved polymerase fidelity in PCR-SSCPA, Mutation Research 303,4 (Dec 1993) 171–175.

[14] Bray, D., Protein molecules as computational elements in living cells, Nature 376 (1995) 307-312.

[15] Britten, Roy J., & Davidson, Eric H.: Repetitive and non-repetitive DNA sequences and a speculation on the origins of evolutionary novelty, Quart. Rev. Biol. 46,2 (1971) 111-133.

[16] Britten, R.J., and Kohne, D.E., Repeated sequences in DNA, Science 161 (1968) 529-540.

[17] Brown, T.A. (ed), Essential Molecular Biology: a practical approach, IRL Press, Oxford, 1991.

[18] Buro, M., A contribution to the determination of Rado's $\Sigma(5)$, Diploma thesis, Aachen, 1990.

[19] Cairns, J., Foster, P.L.,, Adaptive reversion of a frameshift mutation in Escherichia Coli, Genetics 128 (1991) 695–701; Mechanisms of directed mutation, Genetics 131 (1992) 783–789.

[20] Cairns, J., Overbaugh, J, Miller, S., The origin of mutants, Nature 335 (1988) 142–146.

[21] Calderbank, A.R., Hammons, A.R. Jr., Kumar, P. V., Sloane, N.J.A., Solé, P., The $\mathbf{Z}_4$-Linearity of Kerdock, Preparata, Goethals and Related Codes, IEEE Trans. Information Theory 40 (1994) 301–319

[22] Thomas R. Cech, Chromosome end games, Science 266 (Oct 1994) 387–388.

[23] Cech, Thomas R., The chemistry of self-splicing RNA and RNA enzymes, Science 236 (1987) 1532–1539.

[24] Daniel W. Celander and Thomas R. Cech, Visualizing the higher order folding of a catalytic RNA molecule, Science 251 (Jan 1991) 401–407.

[25] Chakrabarti, S., Ranade, A., Yelick, K., Randomized load balancing for tree structured computation, Proc. Scalable High Performance Computing Conf. (1994) 666–673.

[26] Challberg, M.D. and Englund, P.T., The effect of template secondary structure on vaccinia DNA polymerase, J. Biol. Chem. 254,16 (1979) 7820–7826.

[27] Correll, R.A., Feagin, J.E., Riley, G.E., Strickland, T., et al., Trypanosma brucei minicircles encode multiple guide RNAs which can direct editing of extensively overlapping sequences, Nucleic Acids Res. 21,18 (1993) 4313–4320.

[28] Cotton, R.G.H., Detection of single base changes in nucleic acids, Biochem.J. (1989) 1–10.

[29] Crawford, J.and Auton, L.D., Experimental Results on the Cross-Over Point in Satisfiability Problems, Proc. 11th Nat'l Conf. on Artificial Intelligence (AAAI-93) 21–27.

[30] Crowder, H., and Padberg, M.W., Solving large scale symmetric traveling salesman problems to optimality, Management Sci. 26 (1980) 495–509.

[31] Darnell, James, Lodish, Harvey, and Baltimore, David, Molecular cell biology, Scientific American Books, New York, W.H. Freeman 1990.

[32] Deininger, P.L., and Daniels, G.R., The recent evolution of mammalian repetitive segments, Trends Gen. 2 (1986) 76–80.

[33] Dewdney, A.K., Computer recreations column, Scientific American (April 1985) 20–30. A hardware implementation of a Turing machine by G.Uhing found $B_5 \geq 1915$.

[34] Dewdney, A.K., The Turing omnibus, 61 excursions in computer science, Computer Science Press 1989.

[35] Second DIMACS implementation challenge, a book planned to be published by American Mathematical Society, Providence, R.I.

[36] Drake, J.W., A constant rate of spontaneous mutation in DNA-based microbes, Proc. Nat'l. Acad. Sci. USA 88,16 (Aug 1991)

7160–7164. Drake, J.W.: A constant rate of spontaneous mutation in DNA-based microbes, Proc. Nat'l. Acad. Sci. USA 88,16 (Aug 1991) 7160-7164.

[37] Drake, J.W., Rates of spontaneous mutation among RNA viruses, Proc. Nat'l. Acad. Sci. USA 90,9 (May 1993) 4171–4175.

[38] Driver, C.J., McKechnie, S.W., Transposable elements as a factor in the aging of Drosophil

[39] Dubois, O., Andre, P., Boufhkad, Y., Carlier J., Manuscript on SAT algorithm implementations. Should appear in the DIMACS challenge book (qv).

[40] Ellington, A.D. and Szostak, J.W., In vitro selection of RNA molecules that bind specific ligands, Nature 346 (1990) 818–822.

[41] A. R. Fersht, Enzymatic editing mechanisms and the genetic code, Proc. Royal Soc. London B 212 (1981) 351–379.

[42] A. R. Fersht and J.W. Knill-Jones, Fidelity of replication of bacteriophage ϕX174 in vitro and in vivo, J. Molec. Biol. (1983) 633–654. Alan R. Fersht and J.W. Knill-Jones:

[43] Fodor, S.P.A., Rava, R.P., Huang, X.C., Pease, A.C., Holmes, C.P., Adams, C.L., Multiplexed biochemical assays with biological chips, Nature 364 (Aug 1993) 555–556.

[44] Ford, E. and Ares, M. Jr., Synthesis of circular RNA in bacteria and yeast using RNA cyclase ribozymes derived from a group I intron of phage T4. Proc. Nat'l Academy of Sciences USA 91,8 (Apr 1994) 3117–3121.

[45] Foster, P.L., & Trimarchi, J. M., Adaptive reversion of a frameshift mutation in E.Coli by simple base deletions in homopolymeric runs, Science 265 (1994) 407–409.

[46] Fraga, C.G., Shigenaga, M.K., et al., Oxidative damage to DNA during aging: 8-hydroxy-2'-deoxyguanine in rat organ DNA and urine, Proc. Nat'l. Acad. Sci. USA 87 (1990) 4533–4537.

[47] Fredricksen, H., A survey of full length nonlinear shift register cycle algorithms, SIAM Review 24,2 (1982) 195–221.

[48] Garey, Michael R., and Johnson, David S., Computers and intractability : a guide to the theory of NP-completeness, San Francisco: W. H. Freeman, 1979.

[49] Gilbert, W., Why genes in pieces, Nature (News and Views) 271 (1978) 501.

[50] Göringer, H.U., Koslowsky, D.J., Morales, T.H., Stuart, K., The formation of mitochondrial ribonucleoprotein complexes involving guide RNA molecules in Trypanosoma Brucei, Proc. Nat'l. Acad. Sci. USA 91 (Mar 1994) 1776–1780.

bibitem Green, N. Michael, Avidin, Adv. Protein Chem. 29 (1975) 85–133.

[51] Greeve, J, Navaratnam, N, Scott, J., Characterization of the apolipoprotein B mRNA editing enzyme: no similarity to the proposed mechanism of RNA editing in kinetoplastid protozoa, Nucleic Acids Research, 19,13 (1991) 3569–3576.

[52] Guerrier-Takada, C., et al., The RNA moiety of ribonuclease P is the catalytic subunit of the enzyme, Cell 35 (1983) 849-857.

[53] Hagerman, Paul J., Flexibility of DNA, Ann. Rev. Biophys. Chem. 17 (1988) 265–288.

[54] Halford, S.E., Johnson, N.P., The *Eco*RI restriction endonuclease with bacteriophage λ DNA, Biochem. J. 191 (1980) 593-604. See also Clore, Gronenborn, Davies: J. Molec. Biol. 155 (1982) 447–466.

[55] Hjelmfelt, A., Weinberger, E.D., and Ross, J., Chemical implementation of neural networks and Turing machines, Proceeding Nat'l Acad. Sci. USA 88,24 (1991) 10983–10987.

[56] Inoue,T. and Orgel, L.E., A nonenzymatic RNA polymerase model, Science 219 (1983) 859–862.

[57] Jerne, Niels K., The generative grammar of the immune system, Nobel Lecture 8 December 1984.

[58] Joyce, G., RNA evolution and the origins of life, Nature 338 (1989) 217–224.

[59] Joyce, G., Directed molecular evolution, Scientific American (Dec 1992) 90–97.

[60] M. Kálmán et al., Synthesis of a gene for human serum albumin and its expression in Saccharomyces Cerevisiae, Nucleic Acids Research 18,20 (1990) 6075–6081.

[61] Karp, R.M., and Zhang, Y., Randomized parallel algorithms for backtrack search and branch-and-bound computation, J. Assoc. Comput. Machin. 40 (1993) 765–789.

[62] Khorana, H.G., Harvey Lectures 62 (1968) 79. Pure and Appl. Chem 17 (1968) 349. The Biochem. J. 109 (1968) 709.

[63] Kornberg, Arthur, and Baker, Tania A.: DNA Replication, 2nd ed. Freeman 1991.

[64] Kricker, M.C., Drake, J.W., Radman, M., Duplication-targeted DNA methylation and mutagenesis in the evolution of eukaryotic chromosomes, Proc. Nat'l Acad. Sci. USA, 89,3 (1992) 1075–1079.

[65] Kruger, K., et al., Self splicing RNA: Autoexcision and autocyclization of the ribosomal RNA intervening sequence of Tetrahymena, Cell 31 (1982) 147–157.

[66] Kumar, R. & Levings, C.S. III, RNA editing of a chimeric maize mitochondrial gene transcript is sequence specific, Curr. Genet. 23 (1993) 154–159.

[67] Kunkel, T.A., and Eckert, Polymerase chain reaction, p5-10 in: Current Communications in Molecular Biology; CSH Lab Press, Cold Spring Harbor NY (1989).

[68] Lahue, R.S., Au, K.G., Modrich, P., DNA mismatch correction in a defined system. Science, 245 (Jul 1989) 160–164.

[69] Landweber, L.F., and Gilbert, W., RNA editing as a source of genetic variation, Nature 363 (1993) 179–182.

[70] Lewin, B., Genes V, Oxford University Press 1994.

[71] Lindahl, Tomas, Instability and decay of the primary structure of DNA, Nature 362 (Apr 1993) 709–715

[72] Lipton, R., Speeding up computations via molecular biology, manuscript, Dec 9 1994, /ftp/pub/people/rjl/bio.ps on ftp.cs.princeton.edu.

[73] Lombardi, D., Soldati, T., et al., Rab9 functions in transport between late endosomes and the trans Golgi network, Euro. Molec. Biol. Org. Journal 12 (1993) 677–682.

[74] Lonergan, K.M., and Gray, M.W., Editing and transfer RNAs in Acantha amoeba Castellanii Mitochondria, Science 259 (1993) 812–818.

[75] Lorch, J.R., and Szostak, J.W., In vitro evolution of new ribozymes with polynucleotide kinase activity, Nature 371 (Sep 1994) 31–36.

[76] Lundberg, K.S., Shoemaker, D.D., et al., High-fidelity amplification using a thermostable DNA polymerase isolated from Pyrococcus furiosus, Gene 108,1 (Dec 1991) 1–6.

[77] Maniatis, Tom, and Reed, Robin, The role of small nuclear ribonucleoprotein particles in pre-mRNA splicing, Nature 325 (1987) 673–678.

[78] Mantegna, R.N. et al., Linguistic features of noncoding DNA sequences, Phys. Rev. Lett. 73,23 (Dec 1994) 3169–3172.

[79] Markham, A.F., et al., Solid phase phosphotriester synthesis of large oligodeoxyribonucleotides on a polyamide support, Nucleic Acids Research 8, 22 (1980) 5193–5205.

[80] Maslov, Dimitri A., Simpson, Larry, The polarity of editing within a multiple gRNA mediated domain is due to formation of anchors for upstream gRNAs by downstream editing, Cell 70 (1992) 459–467.

[81] Maslov, D.A., et al., Evolution of RNA editing in kinetoplastid protozoa, Nature 368 (Mar 1994) 345–348.

[82] Mathews, Christopher K. & van Holde, K.E., Biochemistry, Benjamin-Cummings 1990.

[83] Mathews, C.K., et al., Bacteriophage T4, Amer. Soc. Microbiologists 1983.

[84] McGinnis, William, and Kuziora, Michael, The molecular architects of body design, Scientific American (Feb 1994) 58–66.

[85] Michel, P., Busy beaver competition and Collatz-like problems, Arch. Math. Logic 32 (1993) 351–367.

[86] Miller, Stanley L. & Orgel, Leslie E., The origins of life on the earth, Englewood Cliffs, N.J., Prentice-Hall 1974.

[87] Minsky, Marvin Lee, Computation: finite and infinite machines, Englewood Cliffs, N.J., Prentice-Hall 1967.

[88] Mitchell, D., Selman, B. and Levesque, H.J., Hard and easy distributions of SAT problems. Proceedings of the Tenth National Conference on Artificial Intelligence (AAAI-92), San Jose, CA (July 1992) 459–465.

[89] Mitchison, G.J., and Wilcox, Michael, Rules governing cell division in Anabaena, Nature 239 (1972) 110–111.

[90] Miyoshi, K-I. et al., Solid phase synthesis of polynucleotides IV, Nucleic Acids Research 8, 22 (1980) 5507–5517.

[91] Monien, B. and Speckenmeyer, E., "Solving satisfiability in less than 2^n steps", Discrete Applied Mathematics, **10** (1985) 287–295.

[92] Mueller, M.W., Hetzer, M., Schweyen, R.J., Group II intron RNA catalysis of progressive nucleotide insertion: a model for RNA editing. Science 261 (Aug 1993) 1035–1038.

[93] Neumann, J. von, and Burks, A.W.: Theory of self reproducing automata, U. Illinois Press 1966.

[94] New England Biolabs catalog (32 Tozer Road, Beverly, MA 01915-5599, USA.)

[95] Nielsen, P., Egholm, M., et al.: Sequence-selective recognition of DNA by strand displacement with a thymine substituted polyamide, Science (Dec 1991) 1497–1500.

[96] Orgel, Leslie E., The origins of life: molecules and natural selection, Wiley 1973.

[97] Orgel, L.E., The origin of life on the earth, Scientific American (Oct 1994) 77–83.

[98] Orgel, L.E., and Crick, F.H.C., Selfish DNA: the ultimate parasite, Nature 284 (1988) 604–607.

[99] Pace, N.R., Origin of life – facing up to physical setting, Cell 65 (1991) 531–533.

[100] Padberg, M.; Rinaldi, G., "Optimization of a 532-city symmetric traveling salesman problem by branch and cut", Operations Research Letters 6,1 (March 1987) 1–7.

[101] Patthy, László, Introns and exons, Current Opinion in structural biology 4 (1994) 383–392.

[102] Pawson, T., Protein molecules and signalling networks, Nature 373 (1995) 573–580.

[103] Peebles, C.L., Perlman, P.S., et al.: A self-splicing RNA excises an intron lariat, Cell 44,2 (Jan 1986) 213–223.

[104] Prescott, D.M., The unusual organization and processing of genomic DNA in hypotrichous cilates, Trends in Genetics 8,12 (Dec 1992) 439–445.

[105] Prescott, D.M., Cutting, Splicing, Reorderding, and elimination of DNA sequences in hypotrichous ciliates, BioEssays 14,5 (May 1992) 317–324.

[106] PROMEGA Biological research products catalog, Promega corp. 2800 Woods hollow road, Madison WI 53711–5399.

[107] Ptashne, Mark: A genetic switch, 2nd ed, Cell Press and Blackwell Scientific Publications 1992.

[108] Roberts, R.J., Restriction and modification enzymes and their recognition sequences, Gene 4,3 (1978) 183-194; Nuc. Acids Res. 16 suppl (1988) r271–r313. Also available through BIONET as <ROBERTS>RESTRICT.{NAR, DOC}.

[109] Robinson, R.M., Minsky's small universal Turing machine, Int'l J. Math 2,5 (1991) 551–562.

[110] Ruth, J., Oligonucleotide with reporter groups attached to the base, chapter 11 in Oligonucleotides and analogs: a practical approach, (F. Eckstein ed.) IRL Press/Oxford U Press, 1991.

[111] Saiki, R.K., et al., Primer directed enzymatic amplification of DNA with thermostable DNA polymerase, Science 239 (1988) 487–491.

[112] Sambrook, J., Fritsch, E.F., and Maniatis, T., Molecular cloning : a laboratory manual (2nd. ed.) Cold Spring Harbor, N.Y. : Cold Spring Harbor Laboratory, 1989.

[113] Sassanfar, M. and Szostak, J.W., An RNA motif that binds ATP, Nature 364 (Aug 1993) 550–553.

[114] Saul, R.L., Ames, B.N., Background levels of DNA damage in the population, Basic Life Sciences 38 (1986) 529–535.

[115] Seiwert, S.D., and Stuart, K., RNA editing: transfer of genetic information from gRNA to precursor mRNA in vitro, Science 266 (Oct 1994) 114–116.

[116] Selman, B.; Levesque, H.; Mitchell, D., A new method for solving hard satisfiability problems. Pages 440-446 in: AAAI-92. Proceedings Tenth National Conference on Artificial Intelligence. (AAAI-92. Proceedings Tenth National Conference on Artificial Intelligence, San Jose, CA, USA, 12-16 July 1992). Menlo Park, CA, AAAI Press, 1992.

[117] Sharp, Philip A., Splicing of messenger RNA precursors, Science 235 (1987) 766–771.

[118] Shore, D., Langowski, J., Baldwin, R.L., DNA flexibility studied by covalent closure of short fragments into circles, Proc. Nat'l Adcad. Sci. USA 78,8 (1981) 4833–4837

[119] Larry Simpson and Dmitri A. Maslov, RNA editing and the evolution of parasites, Science 264 (Jun 1994) 1870–1871.

[120] Simpson, Larry, and Shaw, Janet, RNA editing and the mitochondrial cryptogens of kinetoplastid protozoa, Cell 57 (May 1989) 355–366.

[121] Sogin,M. et al., Phylogenetic meaning of the kingdom concept, an unusual ribosomal DNA from *Giardia lamblia*, Science 243 (1989) 75–77.

[122] Stryer, Lubert, Biochemistry (3rd ed.), Freeman 1988.

[123] Stuart, K., RNA editing in trypanosomatid mitochondria, Annual Rev. Microbiol. 45 (1991) 327–344.

[124] Tausta, S.L., Turner, L.R., Buckley, L.K., Klobutcher, L.A., High fidelity developmental excision of Tec1 transposons and internal eliminated sequences in Euplotes crassus, Nucleic Acids Res. 19, 12 (1991) 3229–3236

[125] Terry,   B.J.,   Jack,   W.E.,   Rubin,   R.A., Mod-rich,P., Thermodynamic parameters governing interaction of EcoRI endonuclease with specific and nonspecific DNA sequences, J.Biol.Chem. 258,16 (1983) 9820–9825.

[126] Tonegawa, Susumu, Somatic generation of immune diversity, Nobel lecture, December 8, 1987.

[127] Tsagris, M, Tabler, M., Sanger, H.L., Ribonuclease T1 generates circular RNA molecules from viroid-specific RNA transcripts by cleavage and intramolecular ligation, Nucleic Acids Research 19, 7 (Apr 1991) 1605-1612.

[128] Uhlenbeck, O.C., A small catalytic oligonucleotide, Nature 328 (1987) 596-600.

[129] J. Watson, Hopkins, Roberts, Steitz, Weiner: Molecular biology of the gene (4th ed), Benjamin-Cummings 1987

[130] B. Weiss et al.,Enzymatic breakage and joining of DNA VI, J.Biol.Chem. 243 (1968) 4543.

[131] Wetmur, J.G., Hybridization and renaturation kinetics of nucleic acids, Ann. Review of Biophysics and Biophysical Chemistry 5 (1976) 337-361.

[132] J.G. Wetmur and N. Davidson, Kinetics of renaturation of DNA, J.Molec. Biol. (1968) 31, 349–370.

[133] Woodruff, R.C., Transposable DNA elements and life history traits. I. Transposition of P DNA elements in somatic cells reduces the lifespan of Drosophila melanogaster. Genetica 86(1-3) (1992) 143–154.

6   Appendix A: Finite state and Turing machine terminology

A *finite state machine* is a finite set of states and finite list of *transition rules*. Each transition rule is of the form "if you are in state '52' and encounter an input 'c', then go to state '31' and output 'g'." The inputs and outputs are optional.

A *Turing machine* (TM) is just like a finite state machine, except that it is equipped with a 1-dimensional "tape" that can be used for storing information. The tape is infinite and is divided into squares, and in each square, a character may be written. The machine is defined by a finite set of rules of the form "if you are in state '76' and the current tape square has the character 'd', then go to state '23', overwrite the current tape square with 'h', and move one square 'left' on the tape." In these rules, the overwriting is optional, and the movement need not be leftward, it could instead be rightward, or no movement at all, depending on the rule.

By convention, Turing machines usually have a special 'halt' state from which no transition is possible, and initially, all but a finite portion of the tape squares are blank. Once the initial message (or "input," or "data and program") has been pre-written on the tape, and you turn on the Turing machine, the rest of its actions are predetermined and no further interaction with a human being is required.

A computer is said to have *universal* computational power over some class $\mathcal{C}$ if it can be programmed to emulate any other kind of computer in $\mathcal{C}$.

Turing machines are universal over the class of *all* machines that would commonly be called computers. Furthermore, there exist particular universal Turing machines (several have been constructed) and any one of them can emulate any other reasonable kind of computer with at most polynomial slowdown. (Really, the present claims are not clearly defined since we haven't said what we mean by "computer," but there are precise emulation theorems involving very large classes of computational devices.) For this reason (loosely speaking) problems insoluble by Turing machines are called Turing-"Undecidable."

A so-called *nondeterministic* Turing machine (NDTM) has a set of transition rules which need not specify a unique action – they can specify two or more contradictory actions. For example, one rule could say to move right, the other could say to move left. In that case, the NDTM splits into two NDTMs, one of which moves left and one of which moves right, and they both continue on from there.

We now speak informally to save time: The class of problems which a universal Turing machine can solve in a number of steps bounded by a polynomial of the number of bits in its input, is called "P." The class of problems which a nondeterministic universal Turing machine can solve (that is, which at least *one* of the alter egos solves) in a number of steps bounded by a polynomial of the number of bits in its input, is called "NP." A problem class X, such that any problem in NP can be rephrased as a problem of the form X (and the rephrasing task is in P) is called "NP-complete." The traveling salesman problem, graph coloring, and a logic problem called "SAT" are known NP-complete problem classes. It is widely conjectured that $P \subset NP$ and that NP-complete problems would require, in the worst case, exponential time to be solved by a (deterministic) TM.

PSPACE is the class of problems which can be solved by TMs which only visit a region of their tape having polynomially large extent. EXPTIME is the class of problems which can be solved by TMs in exponential time. Solving positions of certain games, such as $n \times n$ checkers, is known to be EXPTIME complete. It is widely conjectured that $NP \subset EXPTIME$, so that even NDTMs could not solve such problems in polynomial time. However, a TM which not only could proliferate into several children (ala an NDTM) but also could get information *back* from said children, *could* solve $n \times n$ checkers in polynomial time. This fact illustrates the value of interprocessor communication.

7 Appendix B: Modified DeBruijn sequences

The maximum possible number of letters ℓ in a circular DNA string, so that no k-letter substring occurs twice, is $\ell = A^k$ where $A > 1$ is the alphabet size and $A = 4$ for DNA. Here $\ell \leq A^k$ is obvious, and this bound is met by radix-A DeBruijn sequences, see Frederickson 1982. For noncircular DNA, $\ell = A^k + k - 1$, achieved by noncircular DeBruijn sequences. If we also disallow the appearance of any reversed order and complemented k-letter substring ("modified DeBruijn problem"), then $\ell \leq (A^k + A^{\lfloor k/2 \rfloor})/2 + k - 1$. Conjecturally, this bound is tight when $k = 1$, when $(k, A) = (4, 2)$, when $k = 2$, and for no other (k, A). (The first two facts are easily proven. We have proven the third claim when $2 \leq A \leq 14$ but, suprisingly, we have so far been unable to prove it for all A. The fourth claim is almost wholy conjectural.) Also, conjecturally the upper bound is asymptotically tight when $A \to \infty$ with k fixed (indeed when $k = 3$ and A is even, the formula $\ell = A^3/2 + 2$ seems always to hold!). The usual nonconstructive randomizing argument shows that $\ell \geq k + A^{k/2}$. The best lower bounds that we know for the length ℓ of modified DeBruijn sequences for various (k, A) are given in the table below ("." means "meets the upper bound formula" and "!" means "proven optimal by exhaustive search [by Henry Cejtin].")::

.	k=1	k=2	k=3	k=4	k=5	k=6	k=7	k=8	k=9	k=10	k=11	k=12
A=2	1.	4.	6!	13.	20!	39!	70!	137	264	441	1034	1494
A=3	2.	7.	16!	45	117	315	811	2078	5032	12377		
A=4	2.	11.	34!	133	516	1749	8198					
A=5	3.	16.	64	300	1359	5701						
A=6	3.	22.	110	632	3892	18307						
A=7	4.	29.	172	1106	6869							
A=8	4.	37.	258	1946	16388							
A=7	4.	29.	172	1106	6869							
A=9	5.	46.	360	3019	21831							
A=10	5.	56.	502	4622								
A=11	6.	67.	656	6636								
A=12	6.	79.	866	9496								
A=13	7.	92.	1067	12650								
A=14	7.	106.	1374									

An example of a modified DeBruijn sequence with $A = 4$, $k = 4$, and $\ell = 133$ is

```
abcdabdcccaadddbcdbcacadbaacbccaba
bbdadabbadbcbcbbaabdaabbbbccdbddbd
acbdbdcaabcbaadadbbdbadcdddcdccacd
cbbcdcabddabacbaccbbbddddaadcba.
```

8 Appendix C: DNA basics and terminology

At elevated temperature ($\approx$ 90 C) the two strands which comprise a double-stranded (dsDNA) DNA molecule will dissociate or "denature." The resulting single-stranded DNA (ssDNA) is a linear polymer consisting of a sequence of chemical subunits joined in sequence by strong (covalent) bonds. Each subunit, or "nucleotide," consists of a constant part, a phosphorylated, cyclic, 5-carbon sugar (ribose), attached to a variable part, which is one of four biochemical bases: the purines adenine (A) and guanine (G) or the pyrimidines thymine (T) and cytosine (C). [48]. ssDNA strands have polarity. The monomers have functional groups at two sugar sites, designated $3'$ and $5'$, and during polymerization (below) the monomers are linked together $5' \rightarrow 3'$ in parallel orientation. Ignoring chemical modification to the bases [49] the ssDNA molecule is completely specified by the sequence of its nucleotide bases[50].

While each DNA strand is held together by strong bonds, in solution weak (hydrogen) bonds act to cause *specific* base-pairs to associate: A with T and G with C: these base pairs are "complementary." For any ssDNA sequence there is a complementary sequence consisting of complementary bases written in the opposite order: e.g. the sequence ATTCGCT is complementary to AGCGAAT. The sequences of the two ssDNA strands comprising a piece of dsDNA are complementary, and the strands are held together by the hydrogen bonding between each complementary base. Because it is difficult to seperate two bases without breaking the bonds of their neighbors, the dsDNA binding is highly "cooperative" and the "melting" (dissociation) occurs

---

[48]A note on nomenclature: These are the names of the "bare" bases. Attachment to an un-phosphorylated sugar produces the *nucleosides* adenosine, guanosine, thymidine, and cytidine One, two, or three phosphate groups can be attached to the nucleosides to produce the *nucleotide mono-,di- and tri-phosphates*: eg adenosine 5'-monophosphate. The monophosphate nucleotide forms occuring in DNA are properly referred to as: adenylate, gunaylate, thymidylate, and cytidylate.

[49]Modifications, e.g: methylation, of specific bases to protect DNA sequences from degredation by restriction enzymes, occurs frequently *in vivo*.

[50]Not quite: it is important to keep track of the state of the $3'$ and $5'$ ends, to which either a hydroxyl or phosphate group may be attached, as well. In special cases, it is even possible to attach a $3'$ terminal nucleotide lacking the $3'$ hydroxyl, which blocks further polymerization.

at a defined temperature[51]. When the temperature of the melt is reduced, complementary ssDNA strands re-bind, or "anneal," to form the familiar dsDNA form of DNA.

The distinctive chemical sub-groups of each base which participate in hydrogen-bonding of complementary ssDNA sequences to form dsDNA are buried near the axis of the molecule and not "visible" to the outside. Each base, however, has other distinctive sub-groups by which it can be identified, and these groups are chemically accessible to the exterior of the DNA molecule via the so-called "major" and "minor" groves. A large class of "DNA binding" proteins interact with these exposed base functional groups and bind to specific sequences of dsDNA. Among this class of proteins are the so-called "restriction endonucleases," proteins which bind to specific sub-sequences ("recognition sites") of dsDNA and then cut the covalent bonds linking the individual ssDNA at specific sites relative to the location of the binding sequence.

The covalently-bonded "backbone" of ssDNA, neglecting the bases, is a linear sequence of identical subunits

$$5' - [P - S] - [P - S] - [P - S] - \ldots [P - S] - 3'$$

where $[P - S]$ denotes a (nucleoside monophosphate) monomer consisting of a sugar unit with one attached phosphate group. During polymerization, the reaction between the acidic phosphate group of the incoming monomer and the 3' terminal hydroxyl group of the growing strand is a dehydration reaction much like the reaction of an acid and an alcohol to form an ester plus water. The polymerization of nucleoside monophosphates is, however, thermodynamically unstable with respect to the inverse reaction, hydrolysis: this fact has important implications for the stability of any information-representation involving DNA, discussed below. The polymerization of DNA can be made thermodynamically favorable by choosing monomers consisting of a sugar group attached to three phosphate groups. In the reaction of these nucleoside triphosphate monomers (technically "nucleotides") to form DNA, two of the phosphate groups are split off to produce pyrophosphate $HOP_2O_6^{3-}$. This secondary reaction is sufficiently favored by thermodynamics that the "excess" free energy can be used to "drive" the unfavorable polymerization reaction. A great

---

[51]The melting temperature depends upon sequence, because G-C base-pairing, involving three hydrogen bonds, is tighter than the A-T interaction.

many reactions in biology, similarly unfavorable, are "driven" by coupling to other reactions, very often the hydrolysis of the nucleoside triphosphates ATP or GTP. For the case of DNA polymerization, however, the monomers bring, in effect, the required excess free energy to the reaction, and it is not necessary to couple the reaction to any reaction.

Enzymes known as DNA polymerases catalyze the "template-directed" polymerization of DNA. Provided with an adequate supply of nucleotide triphosphates and ssDNA "template" strands, polymerase binds one of the ssDNA templates and proceeds to "copy" the template to form a complementary strand (which can bind to the template to form dsDNA). Polymerization of the growing strand proceeds by the addition of free nucleotides to the 3' end of the growing strand. As the polymerase can only extend an existing strand, and not initiate a new strand, a short ssDNA "primer" complementary to a subsequence of the template must be added to initate polymerization. Provided with a sufficient supply of primers, cycles of polymerization and denaturation can be used to exponentially "amplify" a population of ssDNA, a technique know as "polymerase chain reaction" or PCR.

The phosphodiester bond is vulnerable to hydrolysis. When hydrolysis breaks ssDNA strands, a "nick" in dsDNA is repairable by the enzyme ligase, provided that the phosphate group remains attached to the 5' sugar. Because the reaction is thermodynamically unfavorable, however, ligase couples the repair reaction to the hydrolysis of exogenous ATP.

Machine-synthesis of ssDNA does not utilize polymerase and requires very different reagents and reactions than occur *in vivo*. In 1980 the yields for each such nucleotide addition (Miyoshi 1980, Markham 1980) were in the 60-95% range, but things have improved since then and $> 99\%$ is now available, enabling the synthesis of oligonucleotide ssDNA sequences with lengths of over 100 nucleotides. The cost of made-to-order DNA is about $2 per base. Methods based on *ligating fragments* which were synthesized in ssDNA form, then converted to dsDNA via polymerase, amplified in bacteria, and selected for correctness, have been used to make a dsDNA 1761 base pairs long, which is the current record (Kalman, 1990). The initial fragments, after cloning, had a $\approx 0.5\%$ error rate per base.

Techniques for the attachment of DNA to magnetic beads, etc., are based on the properties of the bacterial protein streptavidin, which tightly binds to the molecule biotin, a common biochemical cofactor.

Biotin can be covalently linked to a nucleotide base, usually via a short "coupler," while streptavidin can be similarly attached to a substrate or bead; the affinity of streptavidin for biotin then causes the DNA (ss or ds) to bind to the bead. Photosensitive coupling systems are available which debind on exposure to light.

DIMACS Series in Discrete Mathematics
and Theoretical Computer Science
Volume **27**, 1996

Complexity of Restricted and Unrestricted Models of Molecular Computation

Erik Winfree[1]
Computation and Neural Systems
California Institute of Technology
Pasadena, California 91125, USA

`winfree@hope.caltech.edu`

Abstract

In [9] and [2] a formal model for molecular computing was proposed, which makes focused use of affinity purification. The use of PCR was suggested to expand the range of feasible computations, resulting in a second model. In this note, we give a precise characterization of these two models in terms of recognized computational complexity classes, namely branching programs (BP) and nondeterministic branching programs (NBP) respectively. This allows us to give upper and lower bounds on the complexity of desired computations. Examples are given of computable and uncomputable problems, given limited time.

1 Introduction

Molecular computation, as introduced by [1], provides a new approach to solving combinatorial inverse problems, where we are interested in computing $f^{-1}(1)$ for n-bit strings $\underline{x}$ and boolean function f. Instances of NP-complete problems can be expressed in this form; for example 3-SAT. Adleman's technique involves using individual DNA strands to represent potential solution bit-strings $\underline{x}$, then operating on a test tube containing all possible solutions to separate those which satisfy f from those which don't. In many instances, the number of sorting operations required is a low-order polynomial in n, suggesting that

---

[1]This work is supported in part by National Institute for Mental Health (NIMH) Training Grant # 5 T32 MH 19138-05; also by General Motors' Technology Research Partnerships program.

– given exponential space to store the DNA^2 – hard combinatorial problems can be solved efficiently with this technique.

It was not immediately clear, however, what class of boolean functions f could be efficiently inverted. In a clarifying paper by Lipton [9], it was shown that if f can be represented as a size L formula of AND-OR-NOT (AON) operations, then f can be inverted using $2L$ molecular steps using affinity purification[3] only. Lipton suggested further that the use of PCR[4] to duplicate the contents of a test tube would allow an even greater class of functions to be inverted using molecular computation. In this note we follow his program and characterize exactly to what extent PCR helps, in terms of known complexity classes.

As individual steps can take on the order of 15 minutes to an hour, small differences in complexity quickly make the difference between feasible and infeasible experiments. Thus it is of importance to characterize the complexity of these models of molecular computation as carefully as possible. Classes such as "polynomial-size" are too rough to be really useful – we really want to know exactly what polynomial it is.

After defining the two models of molecular computation, we will demonstrate their correspondence with branching programs, and conclude with a few implications of the correspondence.

2   Abstract Models of Molecular Computation

We use the models described in [9] and [2], and use similar notation. These models assume perfect performance of each operation, although in practice the molecular biology techniques are known to be somewhat unreliable. Initial comments on this aspect of the models, and other practical matters, can be found in [2], and will not be address here.

The Restricted Model:

---

[2] In this note we grant that $O(2^n)$ volume is "reasonable". Using substantially more DNA, e.g. to search over additional non-deterministic variables, is considered "cheating". In other words, the question being addressed here is, "Given a fixed amount of DNA, what functions can we easily solve?"

[3] Or some equivalent technique.

[4] Or some equivalent technique.

A *test tube* is a set of molecules of DNA encoding bit-strings of length n. We operate on test tubes as follows:

- *Separate.* Given a tube T and an index[5] i, produce two tubes $+(T, i)$ and $-(T, i)$, where $+(T, i)$ contains all strings where bit i is set, and $-(T, S)$ contains all strings where bit i is cleared. Tube T is destroyed.

- *Merge.* Given tubes T_a and T_b, pour T_b into T_a thereby making $T_a \leftarrow T_a \cup T_b$. Tube T_b is destroyed.

At the end of the computation[6], when we presumably have a single test tube containing all strings in $f^{-1}(1)$, we can use the following operation to sequence the strings $\underline{x}$ in the test tube, as described in [2]:

- *Detect.* Given a tube T, say 'yes' if T contains at least one DNA molecule, and say 'no' if it contains none. Tube T is preserved.

A *program*[7] is a sequence of operations on labelled test tubes. Each statement is of the form:

$$\langle +(T_a, i) \rightarrow T_b; -(T_a, i) \rightarrow T_c; \rangle,$$

where the arrow means "is to be merged with". In other words, one separation and two merges occur for every statement (but note that T_b or T_c may be empty prior to the merge). For clarity, programs can be shown diagrammatically (see Figure 1). At the beginning, all test tubes are empty except for T_1, which contains all 2^n DNA strands encoding all possible input vectors $\underline{x}$. If at the end of the program execution there is a test tube containing exactly those bit strings which satisfy f, then we say say the program has inverted f, or has solved f. The *size* of a program is considered to be the number of statements (here

---

[5]We consider only the case where one variable at a time is tested. More sophisticated operations where multiple DNF minterms are tested simultaneously (see [6]) require more lengthy preparation; thus we argue that the single variable case is not unreasonable for measuring complexity.

[6]We do not consider here whether *Detect* could be used to advantage in the middle of a computation.

[7]The class of programs as given here is slightly different from that given in [2]. In particular, we insist that a labelled test tube is not re-used after its contents have been used (i.e. "destroyed"). The differences are merely a matter of notation, and inconsequential.

Separate operations) in the program. Since programs are considered to be executed sequentially, the size of a program to invert f is often refered to as the time to solve f. The *width* of a program is the maximum number of test tubes co-existing at any given time.

Figure 1:

Implementing an arbitrary symmetric function in $\frac{n(n+1)}{2}$ separations (restricted model).

$$f(\underline{x}) = \text{``}0 < \sum_i x_i < 4\text{''}$$

Given $T_1 = \{0,1\}^n$.

$\langle +(T_1, 1) \to T_3; -(T_1, 1) \to T_2; \rangle$

$\langle +(T_2, 2) \to T_5; -(T_2, 2) \to T_4; \rangle$

$\vdots$

$\langle +(T_9, 4) \to T_T; -(T_9, 4) \to T_T; \rangle$

$\langle +(T_{10}, 4) \to T_F; -(T_{10}, 4) \to T_T; \rangle$

Return T_T.

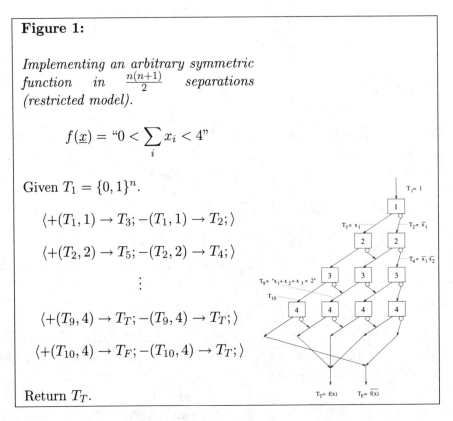

The Unrestricted Model:

The unrestricted model allows one addition type of operation during the computation:

- *Amplify.* Given a tube T produce two tubes T_1 and T_2 with contents identical to T. T is destroyed.

Programs for the unrestricted model consist of statements similar to those for the restricted model, but with the additional form:

$$\langle T_a \to T_b, T_c; \rangle$$

Here the arrow means, "is to be copied into." Unrestricted model programs can also be shown diagrammatically (see Figure 2).

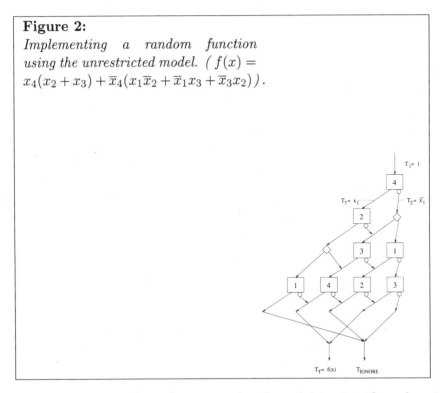

Figure 2:
Implementing a random function using the unrestricted model. ($f(x) = x_4(x_2 + x_3) + \overline{x}_4(x_1\overline{x}_2 + \overline{x}_1x_3 + \overline{x}_3x_2))$.

We might expect that the unrestricted model is siginificantly more powerful than the restricted model. This expectation is quantified and explored in what follows.

3   Branching Programs

Since branching programs are not as familiar a model as formulas, finite-state automata, circuits, Turing machines, etc., it is worthwhile to present an exact definition here. We quote from [16], p. 414:

> A branching program (BP) is a directed acyclic graph consisting of one source (no predecessor), inner nodes of fan-out 2 labelled by Boolean variables and sinks of fan-out 0 labelled by Boolean constants. The computation starts at the source which is also an inner node. If one reaches an inner node labelled by x_i, one proceeds to the left successor, if the i-th input bit a_i equals 0, and one proceeds to the right successor, if a_i equals 1. The BP computes $f \in B_n$[8] if one reaches for the input a a sink labelled by $f(a)$.

---

[8] B_n is the set of all n-input boolean functions.

The size of a BP is the number of inner nodes. Many measures of BP have been studied, especially depth and width.

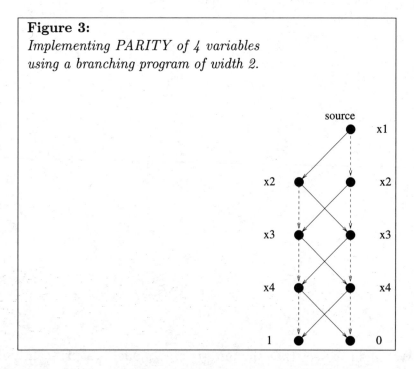

Figure 3:
Implementing PARITY of 4 variables using a branching program of width 2.

We follow [13] in defining a nondeterministic branching program (NBP): we additionally include unlabelled "guessing nodes" of fan-out 2 where both branches are allowed[9]. The NBP computes $f \in B_n$ if by some allowable path one reaches a sink labelled 1 for all $a \in f^{-1}(1)$. The size of an NBP includes the guessing nodes. BP and NBP may be viewed pictorially, as in Figures 3 and 4, in which the designations "left" and "right" are replaced by "dotted-line" and "solid-line" respectively.

---

[9]This definition of NBP coincides exactly with Meinel's 1-time-only nondeterministic branching programs. His more general definitions seem not to be useful in the context of molecular computing.

Figure 4:
Implementing a function using a nondeterministic branching program.
$f(x) =$ *"$\underline{x}$ is palindromic except for isolated (non-adjacent) errors".*
$NBP(f) \leq 2n + 2.$

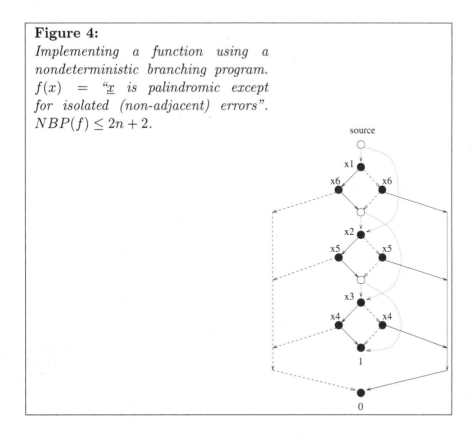

4   Correspondence of Models

Restricted Model $\sim$ Branching Programs

In this section we show that the class of functions which the restricted model can invert in a given time are exactly those functions computed by a branching program of the same size.

Examining Figures 1 and 3, it is clear that not much needs to be proved. The models are essentially identical, except for interpretation. Each separation step corresponds to an inner node of the BP. A strand of DNA corresponds to an input vector for the BP. In summary:

1. If restricted model program P solves f in k steps, then there is a BP G which computes f and is of size k.

2. If BP G computes f and is of size k, then there is a restricted model program P which solves f in k steps.

A single strand of DNA will flow through the test tubes of a restricted model program exactly in the order of inner nodes executed by the associated BP running on an equivalent input vector[10]. Since all possible strands are run in parallel, those that end up in the ouput test tube T_T are exactly the inputs that the BP accepts; i.e. $f^{-1}(1)$.

Unrestricted Model $\sim$ Nondeterministic Branching Programs

In this section we show that the class of functions which the unrestricted model can invert in a given time are exactly those functions computed by a nondeterministic branching program of the same size.

Examining Figures 2 and 4, it is clear that not much needs to be proved. We additionally associate *amplify* statements with guessing nodes in the NBP. Just to be clear, we show:

1. If unrestricted model program P solves f in k steps, then there is a NBP G which computes f and is of size k.

2. If NBP G computes f and is of size k, then there is a unrestricted model program P which solves f in k steps.

We use essentially the same argument as above. However now we say that the set of test tubes which a DNA strand passes through is the same as the set of nodes of the NBP which *could* be activated by the associated input vector. Thus the output test tube contains all strands which *could* cause the NBP to accept; i.e. $f^{-1}(1)$.

5   Corollaries and Conclusions

We now have a theoretical handle on precisely what can and cannot be computed by the restricted and unrestricted models. First, by looking at the polynomial size complexity hierarchy, we can separate the classes of functions solvable by the DNA models.

Many useful results follow immediately from the literature on branching programs. Here is a brief sampler:

- poly-size BP are equivalent to log-space non-uniform TM[11] [11].

---

[10] The author is reminded of some friends who needed to transfer a lot of graphics images from San Francisco to Los Angeles. They considered using ftp over the internet, but on second thought realized it would be faster to put the data in their car and drive, so they did. We are doing the same thing here: We physically move a bunch of DNA through the virtual CPU, one gate at a time – but lots of data simultaneously.

[11] (N)TM = (nondeterministic) Turing machine.

- poly-size NBP are equivalent to log-space non-uniform NTM [11].

- poly-size circuits[12] are equivalent to poly-time non-uniform TM [16].

- thus poly-size BP $\subseteq$ poly-size NBP $\subseteq$ poly-size circuits, where the inclusions are believed to be proper.

- poly-size, constant-width BP are equivalent to log-depth circuits [3] [10].

- $\sqrt[3]{C(f)} \preceq NBP(f) \preceq BP(f) \preceq L(f)$ [13][13].

- $\frac{C(f)}{3} \leq BP(f) \leq L(f) + 1$ [16][14].

With each of these results there is typically an efficient simulation [12]. Other known linear simulations by branching programs include finite-state automata (FSA) and 2-way finite-state automata [3].

As mentioned earlier, results on polynomial equivalence are only of theoretical and not practical relevance. We would like more exact bounds on the complexity of implementing specific functions. The literature on branching programs gives us some such bounds, although admitedly the knowledge is very incomplete. Some known bounds[15] for a few functions[16] are summarized in Table 1.

---

[12]In this note we consider circuits where gates are fan-in 2, arbitrary fan-out, and have arbitrary logic.

[13]C(f) is circuit size, L(f) is AON formula size, etc. $F \preceq G$ means $F = O(G)$.

[14]Note this construction for formulas is better than that given in [9].

[15]See especially [16]: pp. 76, 85, 143, 243, 247, 261, 440; [13]: pp. 50, 51; [8]: pp. 793-797. Note Razborov incorrectly quotes the BP lower bound on MAJORITY [4]. The upper bound comes from [14]. The upper bound on formulas for symmetric functions follows directly from the upper bound Wegener gives for MAJORITY. The upper bound on circuits for DISTINCT comes from a simple application of SORT, followed by adjacent comparisions; a better bound may be achievable. The upper bound on NBP for symmetric functions uses a construction by Lupanov for switching-and-rectifier circuits (see [13]); the construction also works for NBP.

[16]Let $m = \frac{n}{2logn}$, $|\underline{X_i}| = 2logn$ and DISTINCT$(\underline{X_1},\ldots,\underline{X_m}) = 0$ iff $\exists i \neq j$ s.t. $\underline{X_i} = \underline{X_j}$. MAJORITY$(\underline{x}) = 1$ iff $|x| \geq \frac{n}{2}$. PARITY$(\underline{x}) = 1$ iff $|x| \equiv 1$ mod 2. f is SYMMETRIC if f depends only on $|\underline{x}|$, the number of 1's in $\underline{x}$. The lower bounds are for almost all symmetric f.

Table 1. *Lower and upper bounds on complexities under known models for various functions.*

function f_n	$L(f)$	(AON)	$BP(f)$	
PARITY	n^2	n^2	$2n-1$	$2n-1$
DISTINCT	$\Omega(\frac{n^2}{\log n})$	$O(n^2 \log n)$	$\Omega(\frac{n^2}{\log^2 n})$	
MAJORITY	$\Omega(n^2)$	$O(n^{3.37})$	$\Omega(\frac{n \log n}{\log \log n})$	$O(n \log^3 n)$
SYMMETRIC	$\Omega(n \log \log n)$	$O(n^{4.37})$	$\Omega(\frac{n \log n}{\log \log n})$	$O(\frac{n^2}{\log n})$

function f_n	$NBP(f)$		$C(f)$	(B_2)
PARITY		$2n-1$	$n-1$	$n-1$
DISTINCT	$\Omega(\frac{n^{3/2}}{\log n})$		$\Omega(n)$	$O(n \log n)$
MAJORITY	$\Omega(n \log \log \log^* n)$		$\Omega(n)$	$O(n)$
SYMMETRIC		$O(n^{3/2})$	$\Omega(n)$	$O(n)$

6 Discussion

Do we gain anything by using the *amplify* operation? Theoretically, yes, but very little. Contrary to the suggestion in [9][17], we probably cannot invert functions defined by circuits in linear size. Furthermore, in addition to concerns about the reliability of PCR, we should realize that each *amplify* at least doubles the volume of DNA that we have to handle. After just a few such operations, we could practically be unable to continue the computation. For example, if we conclude for practical reasons that 2^{50} molecules of DNA are the most we can handle in one test tube, then we must be very careful not to exceed this limit when merging the products of amplification[18].

The restricted and unrestricted models of molecular computation are still a long way from allowing us to invert algorithmically defined boolean functions. It seems that new molecular operations are necessary if we need this functionality – for example, operations which modify DNA during the computation, such as Adleman's memory

---

[17]It appears that Lipton realized this shortly after distributing his draft. He later characterizes his constructions in terms of *contact networks*, which are related to branching programs (personal communication).

[18]On a similar note, even the restricted model can solve f computed by Meinel's more general NBP model, simply by using 2^m times more DNA volume when there are m non-deterministic variables. This allows computation as efficient as circuits, but at the cost of ridiculous amounts of DNA.

model [2] which can be implemented via site-directed mutagenesis, Beaver's Turing Machine simulation [5] which uses similar mechanisms, or Boneh's *Append* [7], perhaps the simplest and most elegant extension.

Acknowledgments

The author would like to thank Paul W. K. Rothemund, Sam Roweis, and Matthew Cook for their stimulating discussion. Thanks especially to Jehoshua Bruck for pointing me to previous literature on branching programs. Thanks to my advisor John Hopfield for his support and encouragement.

Bibliography

[1] Adleman, Leonard, Molecular computation of solutions to combinatorial problems, *Science* 266:1021–1024 (Nov. 11) 1994.

[2] Adleman, Leonard, On Constructing a Molecular Computer, draft, Jan. 11, 1995.
(ftp://usc.edu/pub/csinfo/papers/adleman/molecular_computer.ps)

[3] Barrington, David A., Bounded Width Branching Programs, PhD Thesis, 1986, Massachusetts Institute of Technology, TR# MIT/LCS/TR-361.

[4] Babai, L., P. Pudlák, V. Rödl, E. Szemeredi, Lower Bounds to the Complexity of Symmetric Boolean Functions, *Theoretical Computer Science* 74 (1990) 313-323.

[5] Beaver, Don, A Universal Molecular Computer, Draft Feb. 6, 1995.
(http://www.cse.psu.edu/~ beaver/research/molec.html)

[6] Boneh, Dan, Christopher Dunworth, and Richard J. Lipton, Breaking DES Using a Molecular Computer, sumbitted to IEEE COMPUTER.
(http:/www.cs.princeton.edu/~ dabo/papers/biocomp.ps)

[7] Boneh, Dan, C. Dunworth, R. Lipton, and J. Sgall, On Computational Power of DNA, to appear.

[8] Boppana, R. B. and M. Sipser, The Complexity of Finite Functions, in *Handbook of Theoretical Computer Science*, ed. J. van Leeuwen, pp. 757-804, 1990, Elsevier Science Publishers B. V.

[9] Lipton, Richard, Speeding up computations via molecular biology, draft Dec. 9, 1994.
(http://www.cs.princeton.edu/~ rjl/bio.ps)

[10] Lipton, Richard, Subquadratic Simulations of Circuits by Branching Programs, in 30th *Annual Symposium on Foundations of Computer Science*, pp. 568–573, 1989, IEEE Computer Society Press.

[11] Meinel, Christoph, *Modified Branching Programs and Their Computational Power, LNCS 370*. 1989, Springer-Verlag.

[12] Pudlák, Pavel, The Hierarchy of Boolean Circuits, *Computers and Artificial Intelligence*, 6 (1987), No. 5, pp. 449–468.

[13] Razborov, Alexander A., Lower Bounds for Deterministic and Nondeterministic Branching Programs, in *Fundamentals of Computation Theory, LNCS 529*, pp. 47–60, 1991, Springer-Verlag.

[14] Sinha, Rakesh Kumar, and Jayram S. Thathachar, Efficient Oblivious Branching Programs for Threshold Functions, in *Proceedings of the 35th Symposium on Foundations of Computer Science*, pp. 309–317, 1994.

[15] Wagner, K. and G. Wechsung, *Computational Complexity*, D. Reidel Publishing Company, 1986.

[16] Wegener, Ingo, *The Complexity of Boolean Functions*, John Wiley & Sons, 1987.

DIMACS Series in Discrete Mathematics
and Theoretical Computer Science
Volume **27**, 1996

On the Computational Power of DNA Annealing and Ligation

Erik Winfree[1]

Computation and Neural Systems

California Institute of Technology

Pasadena, California 91125, USA

winfree@hope.caltech.edu

Abstract

In [20] it was shown that the DNA primitives of *Separate*, *Merge*, and *Amplify* were not sufficiently powerful to invert functions defined by circuits in linear time. Dan Boneh et al [4] show that the addition of a ligation primitive, *Append*, provides the missing power. The question becomes, "How powerful is ligation? Are *Separate*, *Merge*, and *Amplify* necessary at all?" This paper proposes to informally explore the power of annealing and ligation for DNA computation. We conclude, in fact, that annealing and ligation alone are theoretically capable of universal computation.

1   Introduction

When Len Adleman introduced the paradigm of using DNA to solve combinatorial problems [1], his computational scheme involved two distinct phases. To solve the directed Hamiltonian path problem, he first mixed together in a test tube a carefully designed set of DNA oligonucleotide "building blocks", which anneal to each other and are ligated to create long strands of DNA representing paths through the given graph. After this ligation phase, there ensue n steps of affinity purification, whereby exactly the strands representing Hamiltonian paths are separated into a test tube ("the answer").

Richard Lipton [13] subsequently refined the formalism for DNA-based computation. He did away with Adleman's first phase, ligation,

---

[1]This work is supported in part by National Institute for Mental Health (NIMH) Training Grant # 5 T32 MH 19138-05; also by General Motors' Technology Research Partnerships program.

and replaced it by starting all computations with a fixed set of DNA strands representing all n-bit strings. Lipton expanded on Adleman's second phase, separation, where he showed how all solutions to a given boolean formula f can be separated into a test tube ("the answer"). The cost for the generality of this method is indicated by considering solving the Hamiltonian path problem: a straightforward method[2] takes about n^3 separation steps using Lipton's approach, compared to the n steps used by Adleman.

We can conclude from this circumstantial evidence that much of the physical computational power Adleman was exploiting was in his first phase, where annealing and ligation were used. Lipton has explored the power of generalizing Adleman's second phase; we would like now to explore the power of generalizing Adleman's first phase.

An immediate stumbling block is that the chemistry of annealing is not fully understood. At best we can try to define some conditions under which the reactions are predictable, or at least under which it is reasonable to expect that the reactions could be made to be predictable.

2   Some Basic Annealing Reactions

The fundamental chemistry of DNA is based on the double helix and the principle of complementarity. Each strand of DNA is a covalently linked polymer, where each unit consists of a constant part (the sugar-phosphate "backbone") and one of either adenine, thymine, cytosine, or guanine (the bases A, T, C, G). Each strand is oriented; it has a 3' and a 5' end. When DNA forms a double-stranded helix, the strands must be anti-parallel, and complementary bases align (A with T, C with G); such strands are called Watson-Crick complementary sequences. DNA also takes on more complicated configurations, including triple helix,

---

[2]Let the graph have n vertices and e edges; $e \leq n^2$. The best boolean circuit I could devise uses $O(en \log n)$ gates to verify a Hamiltonian path. Another issue is that Adleman's ligation phase requires the synthesis of about $O(n + e)$ oligonucleotides, which is $O(n^2)$ if $e = O(n^2)$; whereas Lipton needs only about $4n \log n$ oligonucleotides to create his standard initial test tube of DNA. However, technology is becoming readily available for synthesizing many oligonucleotides in parallel very quickly (see e.g. [5]); the same cannot be said for the affinity purification steps, which will likely remain expensive. Comparing volume for a graph with $n/2$ edges out of each vertex, Adleman's method uses volume roughly proportional to $(\frac{n}{2})^n$, while Lipton's method uses a volume of $2^{n \log n}$, since it takes $n \log n$ input variable bits to specify a potential path.

quad helix, super-coiled, and branched.

A surprising number of possibilities are available, some of which one may want, and many of which one may not want. DNA is a particularly easy molecule to work with, because it has evolved to be stable, typically unreactive, yet manipulable. RNA and protein, which have evolved to serve many enzymatic functions, are far more reactive, and thus it is less easy to predict how novel designs will behave in an experiment.

I will now comment on some reactions we may wish to exploit, presented in cartoon fashion (Figure 1). I will have to be more detailed with the reactions involved in the main thrust of this paper, where their computation-universality is demonstrated.

(A) This is the canonical annealing reaction for DNA. Two strands with complementary subsequences will form hydrogen bonds and hybridize at the matching base pairs. The rate constants for this reaction, which is reversible, depend on the temperature and salt concentrations, among other things. The melting temperature, above which the complex is not stable, depends upon the number of matching base pairs.

(B) A special case of the above, where the matched region occurs at the ends. Note that the two "sticky ends" (unmatched sequences) are available for further reactions with more DNA.

(C) The above reaction can be used to join two double-stranded DNA molecules with complementary sticky ends. If ligase is present in the solution, the nicks in the backbone of the product will be repaired by the formation of a covalent bond, resulting in two continuous strands.

(D) If mismatches occur flanked by matching regions, the unmatched DNA can bubble out.

(E) As above, except that the mismatch occurs here on both sides. Whether this structure is stable depends critically on the temperature and concentration of salts. For example, a rule of thumb is that the difference in melting temperature between a perfectly matched structure and an imperfectly matched structure is 1 degree per 1% mismatch [19].

(F) This is the simplest DNA branched junction. The assembly of these structures consists of course of sequential steps; only the end product is shown. This 3-armed junction is probably floppy. However, how floppy it is depends upon the exact sequence of base pairs in the oligonucleotides.

(G) This 4-armed junction is commonly known as a Holliday junction. The two horizontal strands tend not to be parallel, but skew. If the sequences along both strands are homologous, then a phenomenon called branch migration can occur, in which the crossover point drifts right or left.

(H) This is the most complicated structure we will consider. We will put it to good use later. It has been found to be fairly rigid and planar [7]. Note the sticky ends. Other related double-crossover junctions are possible, depending upon the number of half-turns present in the helical regions. Ned Seeman calls this molecule "DAE" for double-crossover, antiparallel helical strands, even number of half-turns between crossovers. "DAO", with an odd number of half-turns between the crossovers, has an interesting topological difference: It consists of only 4 strands.

All of the structures above have been made in the lab and their structures verified (see, for example, [7]).

We would ultimately like a theory which could tell us, given a set of oligos, a temperature, and salt concentrations, what stable structures will form, as well as the kinetics. But this is a very complex task!

3   Operations using linear DNA

We will first briefly consider what computations can be performed using annealing and ligation of strictly linear DNA molecules. Many of the possibilities have already been discussed by other authors. For example, the techniques used by Adleman [1] allow for the construction of all DNA representing strings accepted by a finite-state automata (also known as a regular language), using the annealing reactions (B) and (C) above. This is important, because it allows us to create a well-defined, somewhat interesting set of inputs on which to compute in parallel. Beaver has discussed how, in conjunction with polymerase, reactions (D) and (E) can be used to make copies of DNA with context-sensitive insertion, deletion, and replacement of substrings. In light of these powerful operations, it seems plausible that a "one-pot" linear DNA reaction could be designed which performs universal computation.

4   Operations using branched DNA

There are many possibilities for computation using branched DNA. However, since the general chemistry is not well understood, we will try to avoid ungrounded speculation by focusing on one concrete possibility. The rest of this section[3] will concentrate on how to assemble a large "weave" of branched DNA[4] which simulates the operation of a one-dimensional cellular automaton.

4.1   Background: Blocked Cellular Automata

This section develops a formal model of computation called blocked cellular automata[5] (BCA). We will later show how BCA can be simulated by DNA.

The operation of a BCA is diagrammed in Figure 2. As in the Turing Machine model, information is stored in an infinite one-dimensional tape, where each cell contains one of a finite set of symbols. The computation proceeds in steps, where in each step the entire tape is translated, according to a given rule table, into a new tape. The translation occurs locally and in parallel; pairs of two cells are read, and which two symbols are written is governed by look-up in a rule table[6]. It is of critical importance that the reading frame (which cells are paired together) strictly alternates from step to step.

The set of entries $\{(x, y) \rightarrow (u, v)\}$ is called the rule table, or the program, of the BCA. By appropriately designing the rule table, the BCA can be made to perform useful computation. In fact, BCA are computationally universal. A BCA with $k + 3s$ symbols can simulate in linear time the operation of a Turing Machine with k tape symbols

---

[3]The inspiration for this approach comes from the proof of the undecidability of the Tiling Problem (see [10], Chapter 11).

[4]This is, clearly, highly speculative, but we hope not ungrounded.

[5]Blocked cellular automata are a 1D version of what Toffoli and Margolus call partitioning cellular automata in 2D [18].

[6]If the table contains multiple entries for a given pair of read symbols, then the BCA is said to be nondeterministic.

and s head states – the proof is analogous to that in [12]. Thus we can conclude that a BCA can be used to answer any question which can be phrased in terms of a computer program. Small BCA have been designed which sort lists of integers, compute primes, and many other tasks.

A few more comments are in order concerning the abstract model of blocked cellular automata. First we consider the finite-size case. In any attempted implementation of a BCA, we cannot actually construct an infinite tape. Thus boundary conditions become important. We consider the following cases:

(a) No update of boundaries. We start with a finite tape of length $2n$; at each step the tape become 2 cells shorter; and after n steps the computation can proceed no further. This case is not universal.

(b) Inactive boundary conditions. Whenever there is an unpaired cell at either end of the tape, it is copied verbatim onto the new tape. The tape remains always the same size (n cells), and thus there are only k^n possible tapes. As the computation must begin to cycle after k^n steps, this case is also not universal.

(c) Periodic initial conditions. On either side of the input cells we specify a repeating pattern of symbols. Starting with just one copy of the periodic block on either side of the input, computation proceeds as in (a), but if the tape gets too short, we add another copy of the periodic block to either side of the input tape and start the computation anew[7]. This case *is* universal.

(d) Self-regulated boundary conditions. Depending upon what symbol is in the boundary cell, the new tape will either shrink (as in (a)) or expand by appending a new cell to the end of the tape. This case is also universal.

Finally, a word on how an answer is obtained from the BCA. This is a matter of convention. Typically, when the computation is done, the answer is written on the final tape. But how is it known when the computation is done? One possibility is that the tape stops changing; the system has reached a fixed-point. However in this paper we will

---

[7]By memorizing boundary cells, we can avoid re-computing any cells and make the computation more efficient.

consider that a computation is done when a special symbol, called the *halting symbol*, has been written for the first time anywhere on the tape.

4.2   Simulation of BCA by DNA

We will now show how to use DNA to construct a BCA. In this section we will optimistically show what chemical reactions we *hope* will occur; in the following section we consider potential difficulties in finding conditions such that they will in fact occur as we have described.

The DNA representation of the BCA tape is a little counter-intuitive, so we will explain by example. Figure 3 shows part of the DNA molecule encoding the initial tape (the input to the computation). To each tape symbol corresponds a short oligonucleotide sequence, which appears in the initial molecule as a sticky end overhang in the appropriate positions. The rest of the DNA in each segment does not vary with content, and is chosen to maximize structural stability. Note that the reading frame is implicit in the structural form of the DNA. Although Figure 3 is schematic, the 2D picture *is* meant to imply that the whole DNA complex is roughly planar. This is critical, and luckily, it is physically plausible.

There are a variety of ways to make the initial molecule. Note that the initial molecule can be thought of as consisting of several double crossover junctions (from Figure 1H, with the modification that the top and bottom strands are made to be an odd number of half-turns in length – see Figure 6 for detail) linked together by pieces of linear helical DNA. The sticky ends can be designed such that only this unique molecule will self-assemble[8]. Ligase can be added to make the segments of the initial molecule covalently bonded.

We will now explain how the program, that is the rule table, of the BCA is represented in DNA. For each rule, *e.g.* $(x, y) \rightarrow (u, v)$, we create a double crossover molecule whose sticky ends on one helix are $\bar{x}$ and $\bar{y}$, and on the other helix u and v[9] (see Figure 6). All

---

[8]It is easy to see that sticky end sequences can be chosen, using the same techniques as Adleman (see Section 3) , such that a periodic initial molecule will form, creating periodic initial conditions as mentioned in section 4.1 (c) above. Similarly, a regular language of inputs could be made in parallel.

[9]The lengths of all parts of the rule molecules are chosen to be constant for simplicity, but it is conceivable that by using variable length as well as sequence to encode symbols, greater specificity could be achieved.

such rule molecules are added to the solution containing the initial molecule. As shown in Figure 4, what is required for computation is that rule molecules will anneal into position if and only if both sticky ends match.

Eventually, a triangular lattice of linked DNA will form, simulating a triangular region of a BCA corresponding to boundary conditions (a) or (c) in Section 4.1 above (see Figure 5). Boundary conditions (b) and (d) can be simulated by using special rule molecules for the edge of the lattice; the details are not presented here. Note that each level of the lattice has a single strand of DNA which travels the entire length of the lattice at that level, and where the coded symbols occur in the sequence in in which they occur in the BCA at time t.

Finally we ask, how can we access the output of the computation? This breaks down into two questions: How do we know *when* the computation is done? And *what* is on the tape at that point? There are many possible approaches to take; here we will merely sketch one. As mentioned above, we will consider the computation to be done when a special halting symbol is written on the tape[10]. In DNA, this corresponds to the special sticky end motif being incorporated into the lattice. When this occurs, the motif will be present as a double-stranded molecule for the first time, and this site can be be chosen as the recognition domain for a binding protein[11], which could, for example, subsequently catalyze a phosphorescent reaction, turning the solution blue. To determine what is "on the tape" at this point, it is necessary to extract the single strand of DNA corresponding to the final level of the BCA. To do this, first add ligase to covalently bond all the annealed segments[12]. Then add resolvase to break all the crossover junctions[13]. Finally, heat to separate the strands, and use

---

[10]At this point other parts of the tape will typically "not know" that the computation is done, so the lattice will continue to grow. However, it is also possible to design the cellular automaton such that all cells go into a special state to halt computation at the same time (the Firing Squad Problem, see *e.g.* [21]), thereby allowing us to design linear pieces of DNA which fit into the gaps at the final level of the lattice, so that it cannot grow further. This may make extraction of the final tape configuration easier.

[11]The protein must have an active bound form, and inactive unbound form. Furthermore, we must be sure it doesn't bind to rule molecules in the solution.

[12]It is a valid concern that ligase may not be able to bind to any but the outermost strands in a lattice. It may be better to reverse the order of the ligase and resolvase steps.

[13]Although a resolvase has been shown to cut crossovers in double-crossover

affinity purification to extract the strand containing the halting motif. Amplify and sequence that strand however you desire (*e.g.* via PCR and standard sequencing gels).

To summarize the model suggested here, a computation would proceed as follows.

1. First, express your problem via computer program. Convert that program into a (possibly nondeterministic) blocked cellular automaton.

2. Create small molecules (H-shaped and linear) which self-assemble to create the initial molecule (or initial molecules, if search over a FSA-generated set of strings is desired). Add ligase to strengthen the molecule.

3. Create small H-shaped molecules encoding the rule table for your program.

4. Mix the molecules created in steps 2 and 3 together in a test tube, and keep under precise conditions (temperature, salt concentrations) as the DNA lattice crystallizes.

5. When the solution turns blue, ligate, cut the crossovers, and extract the strand with the halting symbol.

6. Sequence the answer.

4.3   Analysis and Estimates. Will it work?

Let's begin the analysis optimistically. The above construction is just one implementation possible in a general class that might be called "crystal computation"[14]. In this class, we design a system where we can tailor-make the energy (and hence free energy) as a function of the configuration. We design it such that the lowest energy state (or in our case, the lowest free-energy state at a given temperature) uniquely

---

molecules [8], it is unknown whether the enzyme will be functional on the inner strands in the lattice. However, the enzyme may be able to, at diminished speed, work from the edges in.

[14]It has been suggested that we shouldn't use the term "crystal", because it has a well-defined special meaning. At best, our constructions yield "pseudo-crystals", because any useful computation is aperiodic. We beg the reader to give us slack in using this term.

represents the answer to our computation. This is closely related to the approach taken by J. J. Hopfield [11] in his seminal work on neural networks. In our case the lowest energy configuration is one where every rule molecule has all four sticky ends bound. Given the presence of the initial molecule, this can only occur if the computation proceeds as desired.

The above analysis is a simplification that fails to take into consideration many aspects of the proposed implementation. For example, it completely ignores the dynamics involved; one simply anneals at a slow enough schedule, the argument goes, and the crystal is the result. Whereas in fact the crystallization proceeds at the edges only, according to kinetics that significantly influence the result.

Can a temperature be found such that two sticky ends bound is stable, while one sticky end bound is unstable? In other words, let T_0, T_1, and T_2 be the melting temperatures for a rule molecule fitting into a lattice slot where respectively 0, 1, and 2 of the sticky end pairings match. We want to keep the test tube at a temperature T such that $T_0 < T_1 < T < T_2$. This should be possible, but how large is the difference between T_1 and T_2? Although this is unknown for the particular molecules we use, we can get some idea by looking at what's known about linear DNA annealing. For example, under standard conditions 20 base-pair oligonucleotides (representing rule molecules with two length 10 sticky ends bound) melt at 70° C, while 14 base-pair oligonucleotides (representing rule molecules with only one length 10 sticky end bound, and the other matching partially) melt at 58° C [19]. $T = 65°$ C would then discriminate the two cases. However, the analogy of rule molecules with two separate binding domains to variable-length oligonucleotides with continuous binding domains is questionable.

A definitive answer to "But will it work?" requires a chemist's knowledge and actual experiments. But we can immediately bring some more concerns to light. Since I do not have answers to them, I will merely mention them in passing. First, to read out an answer of more than one bit, our implementation requires ligating the rule molecules and cutting them with resolvase. It is not at all clear that, in the crowded confines of the DNA lattice, either ligase or resolvase will have room enough to perform its job[15]. Second, it is possible that,

---

[15]If there is an angle between the plane of the lattice and a rule molecule which has just fit in place, then in our construction, an opposite angle is formed when a rule molecule fits into the subsequent layer. Consequently, the 2D lattice, rather than being perfectly planar, folds back and forth like a paper fan, which we call

at a low rate, incorrect rules will be incorporated into the lattice. If this occurs, the computation is ruined. It is thus not clear at this time what yields of correct computation are to be expected, and whether a means could be devised to separate the good from the bad. It is additionally conceivable that stable structures form in the solution unconnected to the initial molecule. For example, four rules molecules could connect in a stable "diamond"; we might think that these complexes will only rarely be formed, because the intermediate steps are unstable (only one sticky end joins molecules), and for similar reasons they would grow slowly. However, they and other types of spurious connections and tangles could form, ruining the computation. A final concern is that there may be some systematic molecular stress or strain that comes into play when building a large crystal, and that beyond a certain size tearing would result. All these issues, and surely others, deserve more attention and study.

If for the moment we suppose that the implementation operates correctly, let us consider what advantage would be derived. Take the following with a bucket of salt: First, a small rule molecule (see Figure 6 for a close-up) consists of 50 base-pairs of DNA, sufficient for sticky ends of length 5, which gives us ≈ 10 symbols[16]. That's 33 K Dalton / rule molecule, with a size probably less than 20 x 44 x 85 Angstroms, for 3 bits / rule molecule.

Assessing speed is even more speculative. Suppose we perform a computation of a 10000-cell BCA with inactive boundary conditions, and compute for 10000 time steps. Suppose it takes 1 second for a rule to fit in when its slot is exposed. Since the 5000 slots are simultaneously exposed, all should be filled in approximately 1 second on average. This leads to a rough estimate of 3 hours for computing the 10000^2 cell lattice. Using 1kg of DNA, we could assemble 10^{19} rule molecules, that is, 10^{11} such calculations in parallel. That leads to a total of 10^{15} operations per second[17]. There is no lab work to be done during this

---

a "corrugated" lattice. The corrugated lattice exposes more of the double helix strands in each rule molecule, possibly making the strands more accessible to ligase but making the crossovers less accessible to resolvase.

[16]We optimistically require only 2 mismatches between sequences representing differing symbols. We also require the complement of a symbol's sequence does not code for a symbol, and that every code sequence has 3 C-G bonds and 2 A-T bonds, for more consistent melting temperatures.

[17]This compares to 300 GFLOPS ($\approx 10^{14}$ basic operations per second) attainable by the best modern supercomputers, e.g. a 7000 processor Intel Paragon. Of course, the "operations" we compare are apples and oranges.

the major stage in the computation. Of course time would also be required in the input and output stages.

4.4   Open questions, extensions, and other speculation.

In addition to the essential question of whether the ideas above can be made to work in the lab, there are many other issues to be investigated.

How energy-efficient is crystal computation?

It is interesting to note that what might be called the computation proper (crystallizing the DNA lattice) theoretically requires no energy at all; in fact, crystallization must be exothermic. Of course, a great deal of energy may be used to heat the mixture up, or to pulse the temperature to dissolve defects. Furthermore, the input and output stages require synthesis and analysis of DNA molecules, and thus also much energy. Our proposal is possibly the most nearly implementable example of the principle that computation is free, but input and ouput are costly [3].

Why use the DAE structure for rule molecules? Clearly the particular choice of molecule is not of intrinsic importance to the idea of this construction. The logical essence is to have an "H"-shaped molecule with four designable sticky ends. At its simplest, one could imagine making the "H" out of two chemically cross-linked strands of DNA (Figure 7a). Another alternative is the slightly larger single crossover Holliday junction. However, it is important for the construction of the lattice that the two linear pieces in the "H" be planar; Holliday junctions have been shown to prefer a (flexible) 60° skew angle [6]. The chemically linked strands imagined above have not yet been characterized. The reason we propose the large double crossover molecules[18] is that they have already been characterized in the lab and are thought to be rigid (which may help prevent tangled lattices) and planar [7]. We chose DAE in preference to other topological variants of double crossover molecules, such as DAO, because the topology of the rule molecule leads to a different "weave" of DNA strands in the lattice (Figure 7bcde). We prefer to have a

---

[18]Ned Seeman suggested we consider double crossover molecules as an improvement over the more awkward branched junction constructions we were originally considering.

single strand which, if covalently linked, runs along an entire level of the lattice, thus encoding the BCA state for that time step.

Why keep around the entire history of the computation? Only the most recent level is necessary for the next step of the computation. *Open question[19]: Can condition be found such that the bottom of the lattice is dissolving while the top of the lattice is growing?* Rule molecules which dissolve at the (hotter?) bottom of the lattice could later be re-used at the (colder?) top.

Automatic programming by evolving rule molecules. Suppose we are interested in finding a small BCA program which generates a particular string, or set of strings. Speculatively, we might begin with a nondeterministic set of all possible rule molecules of a particular size, including some molecules for nondeterministically constructing initial molecules. We grow some 10^{18} lattices, and somehow extract those which compute the desired string. The rule molecules present in these lattices are known to be sufficient to compute the string, but they probably do not contain all possible rules. We now dissolve the "good" lattices and somehow amplify the rule molecules present. Letting lattices grow again, and selecting again for the desired string, we further reduce the nondeterminism of the rule molecules present. We can also consider adding a tiny amount of ligase, thus occasionally creating larger rule molecules from smaller ones – a form of "compiling". Perhaps after a few iterations we look and see what rule molecules are present, or – presuming there is still some nondeterminism – look at what other strings they form. This process is closely related to universal search and can be used, for example, for Kolmogorov complexity based induction[17].

Why a 1D BCA? Why not build a 3D lattice to simulate a 2D BCA? We started with 1D BCA because they can be immediately explored used existing DNA technology. Two dimensions offers several advantages, however, such easier design of efficient computations. Perhaps more importantly, in higher dimensions it becomes easier to design error-tolerant rules [9]; intuitively, point defects in 2D can be filled-in from adjacent correctly-computed cells, while in 1D a point defect severs communication between the left and right side. *Open question: Can the DNA rule molecules be modified so as to build 3D DNA lattices?* Speculatively, one could propose a variant of the double crossover Holliday junction, the "multiple strand double crossover

---

[19]Suggested by Len Adleman, private communication.

junction" (Figure 8), as a means to implement the read-4, write-4 operation required by 2D blocked cellular automata (see e.g [18], Ch. 12). Unfortunately, the proposed building-block molecule has not yet been synthesized.

Potential uses in nano-technology. This paper has suggested an approach to molecular computation via programmable self-assembly. Programmable self-assembly may have other applications. *Open question: Can cellular automata generated lattices be used to define ultra-high resolution electronic circuits?* One possibility, along the lines investigated by Robinson and Seeman [14], would be to conjugate nano-wire onto individual rule molecules, such that when the rule molecules fit together, an electrical circuit is formed. This proposal differs from Robinson and Seeman's suggestion in that whereas they envisioned a periodic lattice of identical memory cells, we suggest that cellular automata rules could be used to build more complicated circuits, either in 2D or 3D.

Why use DNA at all? The principle of computing via crystallization is not restricted to DNA. *Open question[20]: Can non-DNA-based molecules could be used to design desired computations carried out on the surface of a growing crystal?*

5   Comparison with other approaches

Perhaps the most practical suggestion for universal computation via DNA is that of Boneh, Dunworth, Lipton, and Sgall [4]. Their approach makes straightforward use of well understood laboratory techniques for manipulating DNA. They are able to simulate nondeterministic boolean circuits, which seems very efficient for some calculations, and which gives them universal computational ability. Because circuits allow non-local interactions of variable, circuits can be very compact. However, it should be pointed out that the computation requires a lab technician to sequence operations on multiple test tubes; the logic of the program being computed is external to the DNA, which is used as a memory. Small scale computations could be immediately attempted with reasonable chance for success; however due to the weakness of single-stranded DNA and other factors, it is not clear how this approach will scale.

---

[20]Suggested by Stuart Kauffman, private communication.

Other authors have proposed DNA implementations of Turing Machines directly (*e.g.* [2], [16], [15]). The approaches vary from using PCR to relying on restriction enzymes. These approaches show promise, although the reliability and efficiency of the steps is unclear. Furthermore, single-tape, single-head Turing Machines are particularly cumbersome logically; circuits will typically compute the same function in many fewer steps (and single steps take comparable time in both systems – on the order of hours!). In short, although they are of theoretical interest, it is unlikely that anyone will actually go into the lab and solve problems this way.

Our hypothetical cellular automaton implementation differs in a number of ways: First and foremost, our proposal is a "one-pot" reaction. Dump in the rule molecules encoding your problem, and all the logic of the computation is carried out autonomously. No lab work is involved. Furthermore, in addition to running a massive number of computations in parallel, each cellular automaton performs its own computation in parallel – thus fully exploiting the parallelism available. The major and significant drawback of our proposal is that it makes use of chemistry which is not yet fully understood, and thus going into the lab to do a computation this way would be a real technical challenge.

The main conclusion of this paper is that annealing and ligation *alone* may be sufficient for universal "one-pot" DNA computation. Whether the particular scheme envisioned here can be made to work in the lab is a matter for further research. In any case, it is clear that better experimental characterization of the chemistry of annealing is required, and may open up new possibilities for DNA based computation.

Acknowledgments

I would like to thank Paul W. K. Rothemund and Sam Roweis for their stimulating discussion. I am indebted to Ned Seeman for many excellent suggestions, as well as fundamental research on the biochemistry this proposal hopes to exploit; and to Len Adleman for inspiration and great discussions. John Baldeschwieler, Tom Theriault, Marc Unger, Sanjoy Mahajan, Carlos Brody, Dave Kewley, Pam Reinagel, Al Barr, and Stuart Kauffman gave many useful suggestions. Thanks to my advisor John Hopfield for his support and encouragement.

Bibliography

[1] Adleman, Leonard, Molecular computation of solutions to combinatorial problems, *Science* 266:1021-1024 (Nov. 11) 1994.

[2] Beaver, Don, A Universal Molecular Computer, Penn State University Tech Report CSE-95-001

[3] Bennett, Charles H., The Thermodynamics of Computation –a Review, *International Journal of Theoretical Physics* 21 (12):905–940, 1982.

[4] Boneh, Dan, C. Dunworth, R. Lipton, and J. Sgall, On Computational Power of DNA, to appear.

[5] Chetverin, Alexander B., and Fred Russell Kramer, Oligonucleotide Arrays: New Concepts and Possibilities, *Bio / Technology* 12:1093–1099, November 1994.

[6] Eis, P. S., and D. P. Millar, Conformational Distributions of a Four-Way Junction Revealed by Time-Resolved Flourescence Resonance Energy Transfer, *Biochemistry* 32 (50:13852–13860, 1993.

[7] Fu, Tsu-Ju, and Nadrian C. Seeman, DNA Double-Crossover Molecules, *Biochemistry* 32:3211–3220, 1993.

[8] Fu, Tsu-Ju, Börries Kemper, and Nadrian C. Seeman, Cleavage of Double-Crossover Molecules by T4 Endonuclease VII,*Biochemistry* 33:3896–3905, 1994.

[9] Gacs, P., and J. Reif, A Simple Three-Dimensional Real-Time Reliable Cellular Array, *Journal of Computer and System Sciences* 36 (2):125–147, 1988.

[10] Grünbaum, Branko, and G. C. Shephard, *Tilings and Patterns*, W. H. Freeman and Company, New York, 1987.

[11] Hopfield, J. J., Neural Networks and Physical Systems with Emergent Collective Computational Abilities.

[12] Lindgren, K., and M. Nordahl, Universal Computation in Simple One-Dimensional Cellular Automata, *Complex Systems* 4 (3): 299–318, 1990.

[13] Lipton, Richard, Using DNA to Solve NP-complete Problems, *Science* 268:542–545 (Apr. 28) 1995.

[14] Robinson, Bruce H., and Nadrian C. Seeman, The Design of a Biochip: A Self-assembling Molecular-scale Memory Device, *Protein Engineering* 1 (4):295–300, 1987.

[15] Rothemund, P. W. K., A DNA and Restriction Enzyme Implementation of Turing Machines, this volume.

[16] Smith, Warren D., and Allan Schweitzer, DNA Computers in Vitro and Vivo, NECI Technical Report, March 20, 1995.

[17] Solomonoff, Ray, The Applications of Algorithmic Probability to Problems in Artificial Intelligence, In *Uncertainty in Artificial Intelligence*, L. N. Kanal and J. F. Lemmer (editors), Elsevier Science Publishers B. V., North Holland, 1986.

[18] Toffoli, Tommaso, and Norman Margolus, *Cellular Automata Machines*, MIT Press, Cambridge, MA, 1987.

[19] Wetmur, James G., DNA Probes: Applications of the Principles of Nucleic Acid Hybridization, *Critical Reviews in Biochemistry and Molecular Biology* 36 (3/4):227–259, 1991.

[20] Winfree, Erik, Complexity of Restricted and Unrestricted Models of Molecular Computation, this volume.

[21] Yunes, J., Seven-state Solutions to the Firing Squad Synchronization Problem, *Theoretical Computer Science* 127 (2):313–332, 1994.

216　　　　　　　　　　　　　ERIK WINFREE

Figure 1.　　Some basic types of annealing reaction.

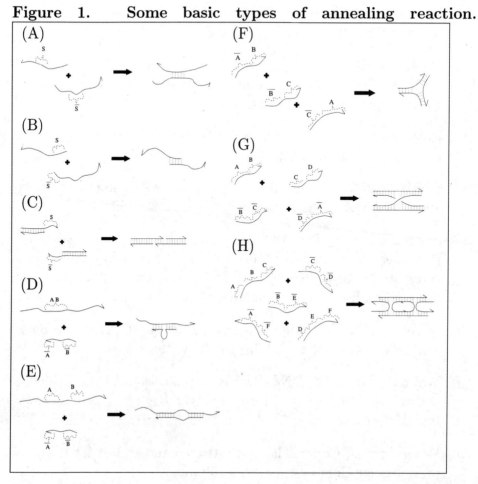

Figure 1: Curves represent single strands of DNA oligonucleotide. The half arrow-head represents the 3' end of the strand. Small lines between strands represent hydrogen bonds joining the strands. The helical structure of the DNA is not represented visually. Letters signify sequence motifs. A bar above a letter signifies the Watson-Crick complement of the motif.

Figure 2. Operation of a blocked cellular automaton.

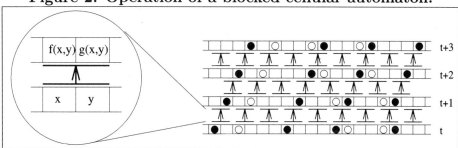

Figure 2: The tape of a BCA, divided into cells, is shown at the bottom right. Each cell contains one of three symbols: blank, black dot, or white dot. The tapes at successive time steps are stacked vertically above the initial tape. The inset, left, details the form of a rule table entry, which governs how new tapes are created.

Figure 3. Encoding the initial tape in a DNA molecule.

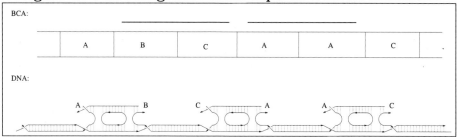

Figure 3: The sequence of sticky ends in the initial molecule encodes the initial tape of the BCA. Thus 'A' denotes a symbol in the BCA diagram, whereas in the DNA diagram it denotes the unique sequence of bases associated with that symbol.

Figure 4. Rule table molecules assemble into the lattice.

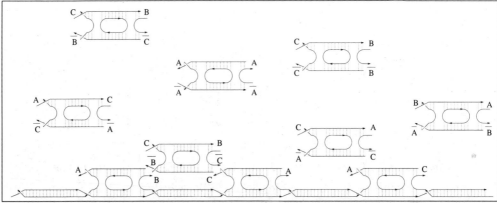

Figure 4: We see free-floating rule table molecules above and the initial molecule at the bottom (both correspond to the BCA in Figure 2). A rule table molecule, with sticky ends $\overline{B}$ and $\overline{C}$, is about to anneal to the initial molecule. At the left, a rule molecule which matches only at $\overline{A}$ will ultimately not stick. Note that the rule molecule with sticky ends $\overline{A}$ and $\overline{A}$ will also not stick, because the orientation of its strands is wrong; this rule molecule will be useful on alternate levels of the lattice.

Figure 5. The DNA lattice resulting from a finite initial molecule.

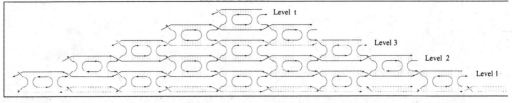

Figure 5: At the chosen annealing temperature, which is above the melting temperature for s base-pair annealing but below the melting temperature for $2s$ base-pair annealing, no more rule molecules can stably attach to this structure. However, if the bottom level (the initial molecule) were extended, then a larger triangle could form. s is the length of the sticky ends in the rule molecules.

Figure 6. Detail of a small rule molecule.

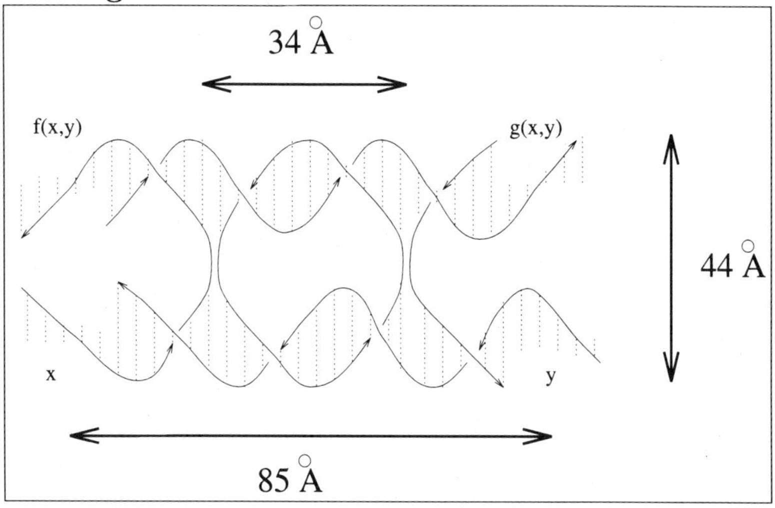

Figure 6: This is the smallest DAE/even style rule molecule possible. It has sticky ends of length 5, and internal region of length 10. Every base pair is shown.

ERIK WINFREE

Figure 7. Alternative Topologies for 2D Lattice.

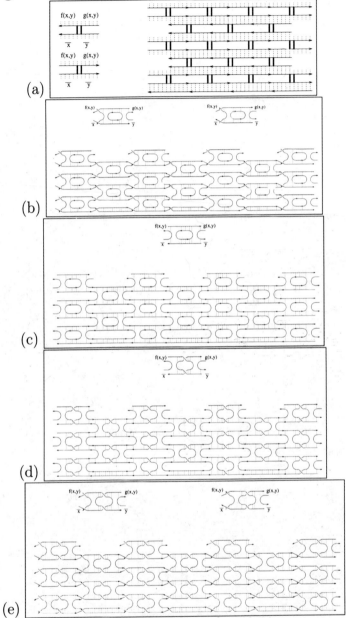

(a)

(b)

(c)

(d)

(e)

Figure 7: (a) Rule molecules based on cross-linked DNA. (b) DAE rule molecules with odd-length spacing. (c) DAE rule molecules with even-length spacing. (d) DAO rule molecules with odd-length spacing. (e) DAO rule molecules with even-length spacing.

Figure 8. A possible 3D lattice of DNA for simulating 2D BCA.

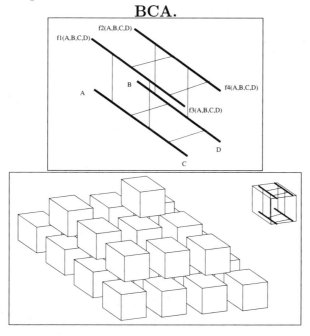

Figure 8: Four DNA double helices may be bound together by crossover junctions (left). Sticky ends determine 2D BCA rules as the rule molecules assemble in an alternative cubic lattice (right).